PERCEPTUAL INTELLIGENCE

may your PI be
sky high!!

Brian Boxer Wachler, M.D.

May you 61 be
Sexy night!!

Michael Gardner, xoxo

Praise for *Perceptual Intelligence*

"Whether you have an intellectual interest in the sciences, psychology, sociology, sports, religion, the creative arts, or sexuality, this book provides brilliant insights and clarity to help you see *yourself* better and identify when your perceptions might be spot-on — or leading you far astray. *Perceptual Intelligence* couldn't be more timely for the chaotic world of today. We're so interconnected through technology but disconnected from each other's perspectives and differing points of view, and that's where being connected *really* counts. The truth is that we introduce our own beliefs, experiences, backgrounds, and DNA into our every perception. So how do we know for sure which perceptions are true and which are distorted by the lens through which we view the world? *Perceptual Intelligence* solves that dilemma and draws you in on every page. Congratulations, Dr. Brian, for speaking with great clarity about things that matter — to people who care."

— **"Dr. Phil" McGraw**, host of *Dr. Phil* and bestselling author of *Life Code*

"*Perceptual Intelligence* will make you reexamine how you see the world and make decisions. Through engaging and often humorous scenarios, Dr. Brian reveals the fundamentals of Perceptual Intelligence, or PI, in ways that can and should resonate with anyone with a curious mind. I see many ways this new understanding of PI can help me make better decisions in my work, home, and finances."

— **Sharon Zezima**, general counsel, GoPro

"A life changer! Not only does Dr. Brian prescribe 'corrective glasses' to see your life more clearly and confidently, he tells you how to create the lenses yourself. I wish I'd had this book decades ago."

— **Leil Lowndes**, bestselling author of *How to Talk to Anyone*

"Perception is everything. *Perceptual Intelligence* shows how reality can be altered in a person's mind concerning so many different things. When I came out as transgender, I had the same sense of purpose I did in my athletic pursuits. I knew that by sharing my journey, I would shed light on the plight of my transgender brothers and sisters, help those who may not know much about the transgender community learn who we are and the challenges we face, and remind them of our shared humanity. That ability to open minds through truth telling is a powerful example of high Perceptual Intelligence."

— **Caitlyn Jenner**, author of *The Secrets of My Life*

"I'm a big believer in Perceptual Intelligence. Without it you won't be successful."

— **Tommy Pham,** Major League Baseball player for St. Louis Cardinals

"Showing us that the more we understand our brains and thinking while competing, the better we perform, *Perceptual Intelligence* will help athletes and nonathletes alike perform better in sports and in life."

— **Pam Shriver,** doubles tennis champion and ESPN commentator

"A must-read for any business leader or marketer who wants to drive more demand for a product or service."

— **Suzanne Saltzman Ginestro,** chief marketing officer for Campbell Fresh, a Division of the Campbell Soup Company

"I found the chapters on the dynamics of social influence and how reciprocity hijacks our perceptual intelligence informative and engagingly written."

— **Robert Cialdini, PhD,** bestselling author of *Influence*

"This uniquely nuanced examination of the power and pitfalls of perception will delight all who seek a greater understanding of how we view the world and the perceptions that influence every aspect of our lives. I will recommend this book to colleagues and patients alike."

— **Andrew Ordon, MD,** "America's plastic surgeon," cohost of *The Doctors,* and author of *Better in 7*

"By discussing how to focus yourself and heighten your PI, Dr. Brian dissects the zone and helps everyone figure out how to get there more often — hence even better performance."

— **Elana Taylor,** Olympic silver and bronze medalist, USA Bobsled

"A masterpiece! Incredibly engaging with well-placed humor throughout, *Perceptual Intelligence* is a clear, grounded, and useful guidebook that teaches the skills we all need to enhance our lives."

— **Lisa Breckenridge,** newscaster for CBS2 in Los Angeles

"Dr. Brian is expanding our view of human perception. Combining a surgeon's precision with the power and fluency of a gifted writer, he challenges us to consider the space between what we understand intellectually and what we feel experientially, the latter of which remains resistant to any deep scientific analysis yet factors greatly in our experience of the world."

— **Khawar Siddique, MD, MBA,** award-winning board-certified neurosurgeon

"Riveting and relevant! At times it boggles the mind — and when was yours last boggled? Read on!"

— **Boze Hadleigh,** author of *Celebrity Feuds!*

"Preparing mentally for races through self-visualization as described by Dr. Brian enhanced our bobsled team's physical athletic abilities and allowed us to achieve our ultimate potential — as individuals and as a team. Before reading *Perceptual Intelligence*, I hadn't thought about applying Perceptual Intelligence to other areas of life, but now I will."

— **Curt Tomasevicz,** Olympic gold and bronze medalist, USA bobsled

"Dr. Brian's new book unlocks so many breakthroughs that anyone who wants to succeed must read it. If you like Malcolm Gladwell's work or just want to see the future before it happens, like in *The Matrix* movies, this is a must-buy book."

— **Justin Hochberg,** creator of *The Profit* and former producer of *The Apprentice*

"The ability to accept all thoughts, fears, grandiose ideas, and doubts as they are and still be able to see and choose a path objectively is the key to success. *Perceptual Intelligence* breaks down that process."

— **Katie Uhlaender,** Skeleton World Cup champion and gold medalist

"Much material has crossed my desk over the years, and I am happy to say that *Perceptual Intelligence* is for *everyone*! Sports, sex, work, and daily living are all explored. Save yourself some time and energy reading about mindfulness in the many books being hyped to you. This one work says it all and will take you to the next level."

— **Bruce Brown,** former head, William Morris Agency, West Coast Literary Department

"Steve Holcomb's legacy was cemented thanks to Dr. Brian, and our team is indebted to him for his incredible work."

— **Steve Mesler,** Olympic gold medalist, USA bobsled

"Having founded and grown multiple successful technology companies, and having been surrounded by some very talented individuals, I was always amazed at how different leaders interpret the same events or facts differently. Dr. Brian's book, *Perceptual Intelligence*, clearly puts the pieces together. I will highly recommend this book to my fellow CEOs."

— **Alex Kazerani,** entrepreneur and former chief technology officer for Verizon Digital Media Services

"A rollicking good read that enhances common sense and awareness, Dr. Brian's book is a tour de force."

— **Dan Routier,** head of sales for Satellites and Orbital Projects, Airbus

"*Perceptual Intelligence* should be required reading for all military officers — on the shelf with Sun Tzu's *The Art of War*. Dr. Brian offers a roadmap to realizing the mind's commanding potential."

— **Lieutenant Colonel Andy Hewitt,** U.S. Marine Corps (Ret.)

"A useful guide both personally and professionally, this book provides a tour of the factors that influence our perception, told with entertaining and straightforward examples, and the practical tools one can use to improve one's own *Perceptual Intelligence*."

— **Jeremy Wacksman,** chief marketing officer, Zillow Group

"Impressive. As a psychological how-to, this book is full of helpful techniques and stories that illuminate, entertain, and surprise. Dr. Brian has a knack for rendering complex medical and scientific support for his theory of *Perceptual Intelligence* in language that is understandable and rewarding and he is a talented storyteller."

— **Thomas Heinkel Miller,** senior lecturer, Department of Communication Studies, UCLA

"Encourages the reader to challenge the way we see the world. Wonderfully written, Dr. Brian's book is a fitting tribute to my friend and teammate, Steven Holcomb."

— **Steve Langton,** Olympic bronze medalist and World Champion, USA bobsled

PERCEPTUAL INTELLIGENCE

The Brain's Secret to
Seeing Past Illusion, Misperception,
and Self-Deception

BRIAN BOXER WACHLER, MD

Foreword by Montel Williams

New World Library
Novato, California

New World Library
14 Pamaron Way
Novato, California 94949

Text design by Tona Pearce Myers

Library of Congress Cataloging-in-Publication Data
Names: Boxer Wachler, Brian S., author.
Title: Perceptual intelligence : the brain's secret to seeing past illusion, misper-
    ception, and self-deception / Brian Boxer Wachler, M.D., Director, Boxer
    Wachler Vision Institute, foreword by Montel Williams.
Description: Novato, California : New World Library, [2017] | Includes bibli-
    ographical references and index.
Identifiers: LCCN 2017019142 (print) | LCCN 2017033891 (ebook) |
    ISBN 9781608684762 (Ebook) | ISBN 9781608684755 (alk. paper)
Subjects: LCSH: Perception.
Classification: LCC BF311 (ebook) | LCC BF311 .B646 2017 (print) | DDC
    121/.34—dc23
LC record available at https://lccn.loc.gov/2017019142

First printing, October 2017
ISBN 978-1-60868-475-5
Ebook ISBN 978-1-60868-476-2

Printed in Canada on 100% postconsumer-waste recycled paper

New World Library is proud to be a Gold Certified Environmentally Responsible Publisher. Publisher certification awarded by Green Press Initiative. www.greenpressinitiative.org

10   9   8   7   6   5   4   3   2   1

This book is dedicated to Steven Holcomb, the record-breaking US Olympic gold medal bobsled driver who unexpectedly passed away in his sleep at the age of thirty-seven while this book was in production (2017). Steven had become a dear family friend over the ten years since I treated his Keratoconus eye disease and restored his vision. He befittingly wore a Superman shirt under his race uniform because he was a true American superhero. This loss left a gaping hole in my family and in the lives of many others. Steven's triumph over his Keratoconus eye disease inspired millions of people and will continue to do so through his legacy. Through the Giving Vision nonprofit organization (www.GivingVision.org), I am committed to ensuring that his legacy of raising awareness about noninvasive treatments for Keratoconus will endure for generations to come.

This book is also dedicated to my mom, Marsha "Bonnie" Wachler (1941–2015), and my dad, Stanley Wachler (who is eighty-five as of this writing). I would not be who I am today without all your love and without your instilling in me the belief that "I can do anything if I believe and work at it." Also, sorry again for forgetting to thank you both during my wedding speech in 1993 (oops!), but thank you for all the haircuts!*

---

* My parents owned a hair salon, and my dad still cuts my hair to this day.

Fair is foul, foul is fair.
— *Macbeth*, WILLIAM SHAKESPEARE

Contents

Foreword by Montel Williams   xv

Introduction   1

Chapter 1. Perception's Seat
The Neurological Underpinnings of Perceptual Intelligence   9

Chapter 2. Mind Over (and Under) Matter
Self-Healing and Self-Sabotage   21

Chapter 3. What You See Is Not What You Get
Mind Tricks and Illusions   31

Chapter 4. Out of Body or Under the Ground
PI and Experiences of Death   45

Chapter 5. Vanity Games
Sand Castles, Card Houses, and the Art of Self-Delusion   59

Chapter 6. Let's Get Physical
PI and Sports   69

Chapter 7. Immaculate Perception
Would You Pay $28K for a Grilled Cheese Sandwich?   87

Chapter 8. The Spell of the Sensuous
How Reciprocity Hijacks Our Perceptual Intelligence   99

Chapter 9. Star Time
Blinded by the Glare of Celebrity   111

Chapter 10. Persexual Intelligence 101
Each to His/Her/Their Own   121

Chapter 11. Gotta Have It
Cat Poop Coffee and the Seduction of Low Production   139

Chapter 12. Are You Different from a Wildebeest in Kenya?
The Dynamics of Social Influence   151

Chapter 13. Fanaticism
The Nature of Extreme Beliefs   165

Chapter 14. The Subjective Experience of Time
And a Chapter Bonus: The Origin of the Bucket List   175

Chapter 15. Gut Check
Following Our Intuition   187

Chapter 16. Your PI Assessment
Can Perceptual Intelligence Be Improved?   199

Epilogue
PI: Your Final Perception   219

Acknowledgments   221

About the Author   225

Notes   229

Index   249

Foreword

It's 1980. I'm twenty-two and just weeks away from graduating from the US Naval Academy in Annapolis, Maryland. I'd made it. I'd come a long way from that very first day four years earlier when I stood with 1,246 other eager, mostly fresh-out-of-high-school kids at this prestigious school on the banks of the confluence of Chesapeake Bay and the Severn River. I would shortly become an officer in the US Navy. The academy isn't for the faint of heart, and being of sound mind and body is a minimum prerequisite for surviving and thriving. It goes without saying that all cadets are in top-flight condition.

I began to have symptoms just before my graduation. For the next two decades they ebbed and flowed. I experienced alternating bouts of pain, weakness, twitching, and vision problems. Doctors diagnosed me with everything from a pinched nerve to an ear infection. My symptoms never stopped me from doing what I wanted, including serving as a naval intelligence officer for twenty-two years. After serving in the navy, I spoke to teens about reaching their full potential. I landed my own daytime talk show, acted in TV shows such as *Jag* and *Touched by an Angel*, and worked as a spokesman for a number of companies. Still, my physical problems never went away. I would lose vision in one of my eyes, only to have it return a few days later. I would experience some numbness or pain, but who doesn't, every now and then, especially if they're physically active?

In 1999 things took a turn for the worse. I had four kids by then, and enough responsibilities for three people. I started experiencing an extreme burning pain in my legs and feet that got so bad that at one point I could barely walk. My doctor ordered an MRI, which finally led to a diagnosis of multiple sclerosis (MS) — a chronic autoimmune disease that damages the myelin sheath, the material that protects nerve fibers, causing neurological transmissions to be slowed or blocked completely, which leads to diminished and sometimes lost functioning.

MS is an unpredictable disease; the symptoms, severity, and duration of MS vary from person to person. Most patients experience muscle weakness and loss of muscular control, fatigue, vision problems, and cognitive impairments such as poor memory and concentration. Other symptoms include pain, tremor, vertigo, bladder and bowel dysfunction, depression, and euphoria. I was diagnosed with relapsing-remitting MS, the most common form of the disease, in which worsening symptoms are followed by periods of remission.

The day of my diagnosis was the worst day of my life. How could I have this illness? It's a disease that women are twice as likely to contract as men. I vividly recall my doctor looking me in the eye and telling me what was in store for me for the rest of my life. He told me that I'd probably be in a wheelchair in less than three years and that I should stop working and exercising, that I should remove anything stressful from my life — which was just about everything. I wondered if my MS was service related or somehow linked to a vaccine I'd received years earlier. Whatever the cause, the news devastated me, and I fell into a deep depression that lasted for months.

But after the shock of having a serious illness wore off, I started taking inventory of my situation. I knew I had a choice to make: either I could spend the rest of my life feeling sorry for myself, or I could view my illness as a call to action. Right there, it became my mission to learn everything I could about MS. I sought out world-renowned doctors, and found some great ones. But it always struck me as odd that my doctor, who didn't know me at all, presumed he could easily determine a plan for the rest of my life. I realized that *I* am responsible for my own health. Today, instead of letting MS control my life, I work to control my disease with healthy eating, exercise, and injections of medication.

The aspect of MS that worried me more than any other was the loss of cognitive functioning. About 34 to 65 percent of folks with MS experience some cognitive decline. One valuable lesson I learned about cognitive functioning and MS has to do with self-perception. What I found both from my own experience and from speaking to therapists — as well as hundreds of MS patients — is that *perceived* cognitive dysfunction has little or even nothing to do with objective decline. In fact, a perceived cognitive deficit, such as memory loss or loss of executive functioning, is highly correlated with emotional distress, including depression. I'm convinced that the ability to exercise some degree of mind control and allay emotional distress can help anyone living with MS more accurately assess his or her own cognitive functioning. For me, improving self-perception has been especially helpful in reducing fatigue — an almost constant companion for anyone with MS. All of us feel tired at one time or another. But if you have MS, then you're in an almost constant state of cognitive and body fatigue.

Shining a positive light on a disease like MS is an ever-evolving process. Every day I carefully monitor my stress level. I've created a vast social support network that includes family, friends, and others who've been instrumental in helping deal with the daily challenges posed by MS.

But I'm convinced that how we perceive our illnesses matters just as much, if not more, in determining the ultimate outcome. And I'm not alone in this belief. Studies have shown that a person's illness perception bears a direct relationship to several important health outcomes, including his or her level of functioning and ability, health care utilization, treatment plan adherence, and even mortality. There's even research suggesting that how we view our illnesses may play a bigger role in determining outcomes than the actual severity of the disease. ·

In general, our illness perceptions emerge from our beliefs about illness and what it means in the context of our lives. While I still have my personal beliefs about the cause of my MS, there's still a lot of uncertainty. Will this be a lifelong struggle? Will a cure be found? How has my illness impacted my family members and friends? Whatever the answers to these questions, I know that it's my perception of MS that will ultimately determine the outcome. I have wonderful doctors, and I'm fortunate to be able to afford the best medical care. But if a therapy or treatment recommendation doesn't fit with my view of my illness, then I often don't stick with it. Or, as one doctor told me candidly, "A treatment that does not consider the patient's view is likely to fail."

It's for these reasons that I jumped at the opportunity to write the foreword for this wonderful book by my good friend and doctor, Brian Boxer Wachler. Dr. Brian has done a masterly job of exploring the underpinnings of human perception, along the way helping us understand our motivations and behavior. As a leading ophthalmologist who performs sight surgeries every day, Dr. Brian has taught me a lot about illness perception and why maintaining a positive outlook is critical to effective treatment and eventual outcome. I know from personal experience that he spends hours with patients, asking questions in order to better understand how they view their illnesses. Dr. Brian was aware of the physical and psychological work I'd done to better cope with my condition. "You have a high degree of Perceptual Intelligence, Montel, because you're unwilling to be done in by your condition. Proactive people perceive any illness, no matter how devastating, as manageable," he told me.

Such conversations have helped Dr. Brian identify patients

who might be at risk of coping poorly with the demands of their condition, and identify and correct any inaccurate beliefs. Once a patient's illness perception has been clearly spelled out, Dr. Brian helps to nudge those beliefs in a direction that is more compatible with treatment or better health outcomes.

Illness perception is a relatively new field. Scientists still don't know how our illness perceptions develop in the first place. And reading a book like *Perceptual Intelligence* could even complicate matters, perhaps making you less sure of yourself. Guess what, folks? That is a good thing.

An important lesson I've learned both from my own life and from Dr. Brian's important book is that the human mind doesn't work the way we think it does. Too many of us hold on to rigid beliefs; we're locked into old paradigms that don't work. We think of memory as objectively truthful, but it's not. Indeed, any memory is likely distorted since it was in part shaped by our perceptions. We perceive cause and effect when accident or correlation is the only thing in play. We see ourselves and the world one way, but in the process we're actually missing out on a lot, not the least of which is the opportunity for self-growth.

Again and again, we think we see and understand the world the way it is, but our perceptions are shaped by experience and our internal dynamics and are therefore beset by everyday illusions and biases. There are myriad ways in which our perceptions deceive us, but this book isn't simply a catalog of human failings. It equally shines a light on our perceptions — where they originate, how they develop, why we so often succumb to them, and what we can even do to increase our Perceptual Intelligence. As Dr. Brian makes clear in this book, we will always have our perceptions, flawed as they might be. The human species will never become completely enlightened about perception, and even if it could, would there ever be a consensus on what "enlightened perceptions" look like? However, it's my hope that by reading this book we will all discover how a better understanding of our perceptions can change the way we think of ourselves and others. Ultimately, *Perceptual Intelligence* provides a way for us to peer into our own consciousness, allowing us to pierce the perceptual veil that clouds our minds and to view the world differently. Wouldn't that be something?

Montel Williams

New York City, 2017

Introduction

Reality is merely an illusion, albeit a very persistent one.
— ALBERT EINSTEIN

In 2009 thousands of visitors flocked to St. Mary's Cathedral in Rathkeale, Ireland, convinced that a gnarled tree stump outside the church depicted a silhouette of the Virgin Mary.

It has been approximated, from various surveys, that 10–25 percent of the world's population claim they have had out-of-body experiences (OBEs).

According to reports from the National UFO Reporting Center, nearly seven thousand unidentified flying objects are reported internationally each year.

Whether or not you are a skeptic, the question you are most likely to ask yourself regarding the above statistics is whether these phenomena are real or merely illusions. Did so many individuals truly have these remarkable experiences firsthand, or were they all reliving Woodstock? If these things did happen, were those who witnessed them interpreting them exactly as they occurred? Or are people having mass hallucinations, possibly even having gone plain mad?

The individuals who have testified to having witnessed the

Virgin Mary in a tree stump, having had an out-of-body experience, or having seen UFOs all have one thing in common: they *passionately believe and are thoroughly convinced that their reality is true.* They are 100 percent certain that what they saw was real, and they are unlikely to be convinced otherwise, even by scientific explanation and logic.

My purpose with this book is not to debunk or even challenge religion, spirituality, or new age phenomena — those fights are all way too big for me to pick. My focus is on Albert Einstein's famed quote. If reality is, in fact, just an illusion, why is it so persistent? Why do our minds so readily accept illusion as reality? More than that, is there a more tangible reality to be found beyond the illusion?

As a surgeon and ophthalmologist who has devoted my career to the field of vision correction, I am fascinated by the connections among our five senses — sight, hearing, smell, taste, and touch — and how our brains register and interpret them to distinguish reality from illusion.

My interest in this subject goes back to when I was a freshman at UCLA in the midst of deciding which classes to take. I had just been turned down by a girl I liked and had gone back to my room to ponder better ways to ask girls out. From a young age, I was fascinated with how the brain worked, especially in the opposite sex. In my elementary school library I secretly checked out *Are You There God? It's Me, Margaret.* As a new college student, I was particularly intrigued by the brain and how information is received and perceptions are formed, so I decided to study psychology as well as biology. Psychobiology became my degree.

Flash-forward to 1999 and my early career as an ophthalmologist. I had a patient, a lifeguard, who was interested in receiving LASIK surgery. On examination, however, I discovered that he suffered from Keratoconus — a disease of the cornea that causes visual distortion — and that it was more advanced in one eye than the other. Yet he hadn't noticed the discrepancy between his eyes. Why? Because his better eye had taken over so much that it had dampened the impact of the blur in the other. Though he was not consciously aware of what was happening with his eyesight, his brain had adapted to this new reality to compensate for what would otherwise be a persistent and annoying sensory overload.

If the brain is capable of physically changing our perception of our own eyesight, I wondered, might it also be altering reality on other matters to suit our psychological needs? For example, if you were to witness an event so painful and emotionally difficult to cope with, might your psyche try to adapt it to best suit your interpretation of the world? Don't our brains — which decipher shades of fact every second of every day — determine what is "real" to help us survive and thrive?

Let's start by putting our cards on the table. Can we say for certain that there is, in fact, a tangible reality? A couch today will still be a couch a year from now, though slightly worse for wear — especially if you have kids and pets and you like pizza. The couch is never going to morph into a dog.

For a select number of people, however, visual misidentification does happen. Folks who suffer from macular degeneration — the loss of sight in the central visual field that can eventually lead to central vision loss — can experience nonpsychiatric hallucinations, since the brain's visual cortex can't properly decode the signals arriving from the eye and compensates by substituting its own fully formed images. The brain doesn't just *fill in* for the thing being looked at but can replace existing images with new ones. Anyone living with Charles Bonnet Syndrome, a condition characterized by visual hallucinations and frequently mistaken as mental illness, is living proof of the maxim that things aren't always as they seem. (We will discuss this at length in chapter 7.) Folks with this syndrome may see anything from abstract patterns to birds and babies and white sandy beaches. However, people who experience these hallucinations know they're illusions, not psychiatric delusions.

The reality of macular degeneration and Charles Bonnet Syndrome raises an interesting question: Is our regular sight also a hallucinated image? If a hundred people see a duck yet one person sees a gerbil, we can be pretty certain that the one person is imagining things (or is just flat-out nuts). But what if 50 percent of the people looking at the duck see the gerbil? Are they all hallucinating, or is there a contagion causing a massive psychiatric illness, such as schizophrenia? Or, as with the visitors who saw the Virgin Mary in a tree stump, is something else going on to cause so many people to see a gerbil when undeniable evidence (such as a photograph) proves that it's a duck?

In this book we will explore the brain's ability to interpret and make sense of the world, why our senses do not always match reality, and how we can influence the world around us through perceptions, inward and outward. Indeed, not everything *is* exactly as it seems, and many factors influence our perceptions. A medical condition known as synesthesia can cause a person to literally see music or taste sounds. (A second form of synesthesia connects objects such as letters and numbers with a sensory perception like color or taste.) Even the common cold, which affects the eyes, ears, nose, and throat — not to mention the brain, when our heads are filled with gook and congestion — has been known to distort our perceptions of everyday events. When we are under the weather from the flu, our perceptions of the world around us might seem so foggy that we may have pessimistic views of occurrences we would ordinarily view optimistically.

As for sleep deprivation, any insomniac or parent of a newborn will tell you that his or her perception of the world is way off track and that memories of the time subsequent to those sleepless nights seem distorted and unreal. And we don't need to review criminal case investigations, forensic evidence, and "beer goggle" studies to see how alcohol and drugs impair our senses and impede our judgment while we're under the influence.

The debate surrounding perception versus reality is a compelling one. We all sense reality through our own perceptual filters. Two people may listen to the same conversation and come away with completely different interpretations of what went down, both equally convinced that they're right. Individuals from rival political parties watch the same televised political debate and have diametrically opposed views of the outcome. A dozen jurors will sit for weeks through a criminal trial, hearing and seeing the exact same evidence and testimony, and yet have a split decision on whether or not the accused is guilty.

How can all these conflicting realities coexist and still be considered "the" reality? We create reality through our own specific sieves. In other words, we *operate in the world we perceive.*

Reality to a peregrine falcon is fiction to a Texas salamander. A red-tailed hawk inhabits a world unimaginable to a star-nosed mole. Humans have a unique take on reality, informed by their upbringing, psychology, biology, genetics, habits, and memories, to name just a few influencing factors. The pope's view of life after death is diametrically opposed to that of the great theoretical physicist Lawrence Krauss. Yet each is convinced that his view is the correct one. Is the pope blinded by faith? Is Dr. Krauss closed to any idea that isn't strictly evidence-based? We all create our own version of the world. It is unlike anyone else's. And how could it not be? It is shaped by *our* perceptions. We generally mold our perceptions like Play-Doh to suit the story we create of our lives. But sometimes our perceptions work behind the scenes, shaping our thoughts and behaviors as if we were asleep at the wheel.

Clearly, our senses connect us to the world. But if our perceptions shape our view of reality, then we're also connected to a manufactured world. Can we ever discern *what is* and *what isn't?*

Many animal species have a view of the world that is inconceivable to us. Whereas humans have a single lens in each eye, insects can have up to twenty-five thousand, depending on the species. (Imagine if we had that many?) What impact do twenty-five thousand simultaneous images have on a fly's perception of the world? Similarly, humans have a flicker fusion threshold — the frequency at which an intermittent light stimulus appears to be completely steady — of fifty images per second. Anything slower is captured one image at a time; anything faster appears as continuous motion. Chickens, on the other hand, don't see continuous motion until the speed reaches a hundred flickers per second, and flies don't see it until three hundred flickers per second. For these animals, the world doesn't turn into a movie until we're already on our second bag of popcorn.

How do we humans make sense of our experiences? How does the human mind move beyond purely reactionary reflections of what it encounters? Unlike sharks, we will never be able to detect the faint, distant electrical impulses released by a dying fish or smell a drop of blood in a million drops of water a quarter mile away. We do have one advantage: our ability to reason. It is our barrier between illusion and reality.

What Is Perceptual Intelligence?

I've seen different definitions of Perceptual Intelligence (PI), but I like to think of it as *how we interpret and occasionally manipulate our experiences to distinguish fantasy from reality.* PI relies heavily on our senses and instincts, but it is frequently influenced and distorted by our emotions and memories. Just as with other forms of intelligence, some people have higher PI than others. However, PI is an acquired skill. It begins with awareness and requires practice before it becomes habitual. So you may find yourself initially overreacting to a situation or circumstance, but with proper knowledge and a different perspective, you may start to ask yourself: Am I interpreting the situation correctly and making the best possible choice?

In their excellent book *The User's Manual for the Brain,* authors L. Michael Hall and Bob G. Bodenhamer write, "The problem is never the person, never the experience, never what we have been through. The problem is always the frame, always the mental movie, always the higher frames running the movie." It's how we interpret what happens to us. If a bird with good aim uses my head for target practice, I could either get ticked off or say, "It's good luck!" (which I learned from my Brooklyn-born dad).

When we have a vague memory of a painful incident, what purpose does it serve? Why bother keeping that potentially incorrect perception of the event when you can make something good come of it? This is where the sniper ability of Perceptual Intelligence comes in. A well-developed PI can identify and take down a faulty idea that tries to sabotage you. Having high PI is recognizing that your mind is more plastic than you think and can be molded and reworked as needed. PI can be improved, just like any other skill, such as driving a car, playing a sport, or learning an instrument.

Many people have survived traumatic incidents and made life decisions based on these experiences. Their perspectives on these events shaped their lives in either a positive or negative direction. It was not the incidents themselves that determined the outcomes; it was the individuals' *perceptions* of the incidents and how they reacted afterward that formed their future. The "heroic" survivors we see on TV or read about in books merely applied principles of PI, whereas the "victims" remained immobilized.

Sharpening Your Perceptual Intelligence

As I've said, since PI is a learned skill, it requires practice before it becomes a habit. Someday you may find yourself initially reacting unfavorably to a situation. Rather than jumping to a negative interpretation, you can catch yourself and ask: Is this the best choice? If not, you can tap into your PI, change your perspective, and achieve a more favorable outcome.

My main objective in writing this book is to help you find your *aha!* moment regarding how you perceive and react to the world from the inside out. It is my sincere hope that, as you follow me on this journey, you will discover a heightened and enlightened understanding of the mystery that is human perception and make better decisions based on what your senses and intuition are telling you. In the chapters that follow I will reveal to you how understanding and maximizing PI is the key to what lurks behind your thoughts, behaviors, and feelings. You'll learn about (not necessarily in this order):

- Why some people can't resist "cat poop coffee" at $100 a cup
- How the brain helps us make sense of the world

- When the mind is healing us and when it's doing more harm than good
- Why we hold on to our illusions
- Why we feel compelled to "return a favor"
- What is really happening when we see aliens in the middle of the night
- How low PI helps us enjoy art more
- Why some people see Jesus in their cornflakes
- How inflated PI can influence the PI of the masses
- Why some athletes and teams are winners and others chronic losers
- How reciprocity hijacks our perceptions
- How celebrity personas manipulate us
- How Mark Twain's thoughts on masturbation reveal his high PI
- Why Kim Kardashian West has so much social impact
- How cults brainwash people with low PI
- Why our perceptions of time are so often distorted
- When to listen to your gut

Throughout the book I've included examples and case studies of people with extraordinary PI and some other examples that might cause you to spit out your coffee. Last, in chapter 16, you'll find an assessment that will help you determine if you have low or high PI, as well as a few simple methods for increasing your PI.

A hundred years from now, we may not even recognize the science that is being practiced today. We will almost certainly have completely mapped the brain by then, yet we may still be no closer to understanding how we perceive the world. In the future, as with today, we will continue to perceive. Somehow, every day we will convert the inconceivable into the conceivable, as do all living things in their own inimitable way. As you'll see in the next chapter, it all starts with the human brain.

Fasten your seatbelt and brace for impact.

Perception's Seat

The Neurological Underpinnings
of Perceptual Intelligence

You're walking alone through a forest late on a windy night. You've been through these woods many times before and know the way, but you've never done so by yourself, and certainly never after dark. You arrive at a clearing and see something unusual in the distance. It blocks your path, so you proceed cautiously while squinting in the darkness to determine what lurks there. Suddenly, you freeze as you see the form take the shape of a large animal; its skin glistens in the moonlight. You can feel its glowing eyes bearing down on you, sizing you up and preparing to attack. Its razor-sharp fangs jut out. It approaches you, and your heart races while you debate whether your action will be fight or flight. As you are about to make a run for it, the creature lunges right at you and your reflexes kick in to protect your face. You scream at the top of your lungs until you realize that you have been scared out of your wits by a plastic garbage bag full of leaves and twigs that had been caught in a gust of wind. Brushing off the refuse, you laugh at yourself for having thought this stupid garbage bag was a monstrous animal. You continue on your way.

In this scenario, at what stage did your perception of the animal originate? Were you expecting to see something dangerous because you were alone in this dark forest for the first time? When the object first appeared in your line of vision, did you tense up and become fearful?

Misperceiving the bag reflects poor PI in that situation. From the beginning, the menacing object was merely a plastic bag full of leaves and twigs. It was harmless all along, but in the absence of light you couldn't discern what it was; then your imagination kicked in, filling in the missing details. Innumerable memories might have kicked in, presenting subliminal messaging: perhaps you had seen a horror film in which someone was attacked by a hideous animal in the woods; you may have read something in the news about a predator on the loose in that forest; when you were a child, perhaps your mother warned you about never going into the forest because she had a relative who had been harmed there by a wild animal; or, as is quite common, the forest evoked sinister images from fairy tales like "Little Red Riding Hood" or literary classics like *Sleepy Hollow.*

Every individual perceives things differently. The same circumstances may produce wildly different interpretations, depending on one's PI. Another person might have been in the same situation as you but determined right away that the bag was a harmless object taking the shape of an animal. On the other hand, a third person venturing forth might have panicked and closed his or her eyes the entire time out of fright. The plastic bag and its contents would have hit his or her unprotected face dead-on, causing scratches and bruises. This terrified individual would scurry off and post a story on the Internet about having been attacked by a supernatural monster.

In this chapter we will explore the world of human perception: what it means, how and where it originates, and how the brain functions and tries to make sense of it all.

Are We All Trapped in a Matrix?

In 1999 Andy and Larry Wachowski created an enormously successful science fiction film called *The Matrix*, which starred Keanu

Reeves as Neo, a man who discovers that the world we live in isn't real. The human race is actually part of a vast computer simulation in which people's energy is being used to power machines ruling the "real world." While in the Matrix, everything humans experience is convincingly real; they see, feel, touch, taste, and smell everything. Their memories and emotions all originate from what occurs in the Matrix, and this has been their perception of reality for years... until Neo joins a rebellion and battles to free the minds and bodies of humanity to join the real world, no matter how dark and precarious it might be.

The film's executive producer Andrew Mason distilled the movie down to this: "[It] is the question of whether or not what I am experiencing right now is real." *The Matrix*, of course, is just science fiction. (I hope — how would we prove otherwise?) But the notion that we might not have control over our perceptions or that we aren't living in the "true reality" is a fascinating one. As with the scenario of the bag of leaves, can we be so easily fooled?

Our brains are obviously our primary "matrix" and vital to our perception of the world around us at all times. The rest of our body sends inputs to the brain — from pain to pleasure and everything in between — which in turn responds with interpretations of them; not only does it tell us if and how we should react, but it stores the information for later recovery and analysis.

The brain is an organ, though it is often called a muscle because it "performs work" and rules all other muscles from its perch in the skull. The brain can be dissected, measured, and studied. But there's also something we refer to as the "mind," which gives rise to our consciousness. Unlike the brain, the mind is transcendent and nonquantifiable. Neuroscientist Sam Harris describes it as "what it's like to be you." Unlike a kidney, heart, or lung, you can't transplant consciousness from one person into another.

The vexing question is: What role does the *mind* play in perception and PI as opposed to that played by the brain? Is the mind the seat of perception within the brain itself? Perhaps it is microscopic or even invisible and located within some lobe or synapse? Or does it reside somewhere else?

These questions are further complicated when we add the body to the equation. Sometimes it seems the body is just the brain's personal chauffeur. But recent studies in neuroscience argue convincingly that the brain and nervous system are so intertwined via a complex web of mutual receptors and connections that it's absurd to speak of them as separate entities, as in the Cartesian view of the body. Thus, any explanation of Perceptual Intelligence hinges on the neurobiological argument that while mind and body are interdependent, the mind is not the brain. By demonstrating the truth of this assertion, we can perhaps identify the source of PI — without any paranoia of being trapped in a matrix that is directing the entire cosmic show.

Opening the Doors of Perception

In the introduction I defined Perceptual Intelligence as *how we interpret and occasionally manipulate our experiences to distinguish fantasy from reality.*

But what, exactly, does the word *perception* mean in this context?

The Merriam-Webster online dictionary defines *perception* as "the way you think about or understand someone or something...the ability to understand or notice something easily... the way you notice or understand something using one of your senses." This covers it on a basic level, but it doesn't begin to explain the role perception plays when *interpretation* is based on myriad factors outside our senses — namely, intuition (or gut feeling), personal experiences, timing, and so forth.

Interpreting what we experience, therefore, requires something much greater than perception alone. When we perceive something — such as the scary object in the Sleepy Hollow forest — it doesn't mean that we are in any way achieving accuracy.

Perception involves immediate, raw, and unfiltered data that enters our minds before being processed through thought and action. Once we interpret the object as having glistening skin, glowing eyes, and razor-sharp fangs, our minds have taken an extraordinary leap; our perception has turned our conclusion into our new reality. We would get an F for our Perceptual Intelligence in this scenario.

In his book *An Inquiry into the Human Mind: On the Principles of Common Sense,* eighteenth-century Scottish philosopher Thomas Reid examined the concept of immediacy, arguing that the perception process had to include some idea or notion of the perceived object: "If, therefore, we attend to that act of our mind which we call the perception of an external object of sense, we shall find it in these three things: First, some conception or notion of the object perceived; Secondly, a strong and irresistible conviction and belief of its present existence; and, Thirdly, that this conviction and belief are immediate, and not the effect of reasoning."

Even as babies, we sense and recognize objects in their immediate form, allowing us to create an immediate, unmediated experience. In this way we convince ourselves of the reality of a perceived object. Once this perception links with sensation — a direct function of the brain — the experience of the senses creates a link with the brain's mental models.

Think about smelling a rose. While the fragrance is merely a perception, it becomes a sensation when we realize that the rose smells good — unless you happen to be allergic to roses. Smell can be a vehicle for our sensing pleasure or for our annoyance. Sensation relies on the mere act of smelling, whereas perception relies on interpretation of that aroma. If you believe you will see a rose next week, you already anticipate a certain pleasing smell; if you happen to be allergic to roses, you would head down a different path to avert a visit to the allergist. These two variant responses are the basic principles of PI in motion.

IF A TREE FALLS,
OR HOW TO GIVE YOURSELF A PHILOSOPHICAL HEADACHE

It's been a philosophical and scientific puzzle for centuries, dating back to 1710 and Irish philosopher George Berkeley: *If a tree falls and no one is around to hear it, does it still make a sound?* The answer is unequivocally *yes* — but that's because I've tricked you. Sorry about that, but I need to make a point: "no one" implies human beings; certainly animals are capable of hearing, so the sound of the tree falling is certainly detectable by those nonhuman creatures within earshot. But does a sound need to be physically perceived by a *living being* in order to be real?

Philosophers continue to debate this — especially those fixated on the concept that our senses exist only in our minds — but many physicists look toward quantum mechanics theory for the answer. Sound is produced when one source causes the molecules of another (such as air or water) to vibrate, producing a molecular wave. So the technical answer, from a scientific perspective, is yes; any tree that falls will always produce a sound, whether or not we (or nearby animals) are there to register it, because the molecular wave has been produced.

Let's take the scientific perspective a step further. We have a general idea of what a tree falling sounds like: *thud.* If no human was there to hear the thud, we still know it occurred because our memory of the sound — which we perceived as a reality at the time from our sense of hearing and perhaps from simultaneously feeling the vibration — has kicked in. Our mind has interpreted the molecular wave as a thud because we have come to know the sound a tree makes when it strikes the ground from a variety of potential sources, such as having heard it in the woods or having heard a recording of it (in films, on television, or on the radio). We have never formulated a scenario whereby the tree does *not* make the thud sound.

Sensory Overload

Sight, sound, smell, taste, and touch: each sense is important enough to warrant being overseen by its own dedicated sensory organ (eyes, ears, nose, mouth, and skin). Our world is teeming with stimuli, and our senses react to them in both voluntary and involuntary ways. We might happen to inhale the scent of a rose along a path or we might spot the flower, deliberately step toward it, press out nostrils against it, and breathe in the fragrance to our hearts' content.

No doubt you are wondering how this process actually works. It requires a complex interplay of one or more senses operating either simultaneously or separately to make something perceivable. Our sensory organs contain receptor cells that detect physical sensations. There are general receptors, which are found throughout the body; special receptors that include chemical receptors (chemoreceptors) in the nose and mouth; photoreceptors in the eyes; and mechanoreceptors in the ears. These receptor cells register the stimulus, converting its energy into an electrochemical signal that relays information about the stimulus through the nervous system to the brain.

From there, these electrical signs are conducted to a nearby area of primary processing, where the initial characteristics of the information are elaborated, according to the original stimulus's nature — that is, its smell, taste, feel, and so on. Then the already modified information is transmitted to the thalamus, a structure lying deep in the brain that is involved in sensory and motor signal relay and the regulation of consciousness and sleep.

The thalamus plays the indispensable role of primary gatekeeper for our senses, determining which signals head to the cerebral cortex. In order for us to see, for example, the retina must send an input via the optic nerves to the thalamus; this is where older data connects to the new information to form a message, which is then transported to the visual cortex of the brain.

Let us not forget the cerebral neocortex, a critical part of the cerebral cortex that helps control functions such as sensory perception, motor commands, spatial reasoning, conscious thought, and, in humans, language. Some studies indicate that it's the neocortex that is responsible for the differences in how we experience the world around us.

It all sounds pretty simple, right? But there is so much more to our brains than meets the eye.

DOES THE BRAIN MAKE PERCEPTION, OR DOES PERCEPTION MAKE THE BRAIN?

Here's something to consider: Almost everything we perceive might be nothing more than an internal mental simulation of the external world, informed occasionally by sampling data culled from our senses. Now, before you raise your eyebrows in disbelief, realize this isn't some in vogue new age theory but a widely held scientific concept.

In other words, we are right smack in the center of the Matrix (though we are not provided a false reality by machines). Reality is real, but what we see, hear, feel, touch, and smell is all in our heads, courtesy of our brain's "built-in virtual reality machine," as described by Bahar Gholipour, managing director of BrainDecoder.com. There are, of course, many neuroscientists who might disagree with this notion.

Whichever the case may be, brain scientists seem to agree that both sensory information and our mental models play a critical role in how we make sense of the world. From the first moments of life, our brains learn from experience and construct images as a way of predicting future interactions with the environment.

This is a primitive survival game. Our brains simply can't process billions of sensory inputs in minute detail, so we tap into our previous experiences to fill in the blanks, speed up the process, or jump to emotionally driven conclusions. You assume that the red flashing lights in your rearview mirror are those of a police car because you just rolled through a stop sign. Your heart begins to race, since you got a ticket two months ago and your insurance rates will go up with this ticket. It would take too long to break down all the sensory inputs of that scenario, but suffice it to say your mind misled you and reduced your PI. The red flashing lights belonged to an ambulance. Your PI is most certainly not bulletproof.

Since we are consciously and subconsciously making judgments on so much data — many times without our senses having performed a fresh or complete survey — it's reasonable to believe that as much as 90 percent of human perception is converted to mental fabrication, though by luck we still occasionally happen upon an accurate conclusion (or come close enough). Our ability to recognize how and where our brains have tampered with our perceptions is where we can harness our PI and control our sense of reality.

Brain and Brain — What Is Brain?

You might think we know a lot about the human brain. After all, brain research is surging. In the past two years, both the United States and the European Union launched new programs to better understand it. Technology for recording brain activity has been improving at a revolutionary pace. Scientists are finally starting to grasp the overwhelming complexity of illnesses that afflict the brain and how to treat these illnesses. At Ohio State University, scientists have used skin cells to grow a "mini brain," or *organoid,* that is the genetic equivalent of a human fetal brain and could be used to fight cancer and autism, as well as Parkinson's, and Alzheimer's, and a number of other debilitating neurological ailments. All this comes on the heels of an ambitious Brain Research through Advancing Innovative Neurotechnologies (BRAIN) Initiative — an acronym that no doubt made some scientist feel clever — designed to speed "our understanding of the brain at the level of its neural circuitry" and "develop a fundamental understanding of the brain."

Yet, as the *New York Times* reported in 2014, the growing body of information about the brain exemplifies a paradox of progress: all these great advances simply serve to highlight how little we actually know. As Aristotle opined, "the more you know, the more you know you don't know." It's the understatement of understatements to declare that the brain is complex. The average human brain has about 90 billion neurons that make 100 trillion connections or synapses. With anything this complicated there will always remain unanswered questions.

The neurological breakdown I provided earlier in this chapter explains the boiled-down A to B to C of brain perception from a technical standpoint and helps our understanding immensely from a medical perspective; health practitioners identify and treat illnesses affecting these systems every day. Yet there is so much we still don't know and understand about the *hows* and *whys* of brain perception and the stops between A, B, and C. The brain is a neurological "peekaboo."

Although our brains are constantly interpreting inputs, we are not computers or robots. In fact, there is nothing concrete about how our brains handle the incessant flow of messaging; they experience electromagnetic waves not as actual waves but as images and colors. When you look at your cell phone, you don't consciously see the wavelengths of all the different colors on your screen. (Or at least I hope you don't — if you do, I'm sure the military would like to detain you for some in-depth probing.) When you listen to a song, you don't smile inside because of the pleasing amplitudes and frequencies of those sounds. When you walk into a restaurant and catch a whiff of what the chef is preparing in the kitchen, do you think, "Wow! Those chemical compounds dissolved in the air that reached my olfactory sense sure are pleasing!"? Of course not. Instead, you look forward to tucking in to a great meal. Our perceptions take shortcuts because doing so is more efficient. Colors, sounds, smells, and tastes are the end products of our sensory experiences.

Our brain has a seemingly impossible job: incorporating external stimuli with what's self-generated. This may not seem like any great shakes, but it helps illuminate how we draw a distinction between the world and ourselves. It's the very beginning of how a brain sorts through the data tsunami coming in from the outside world and how it ascribes meaning to it.

That is part of the brain's job, after all — to build an image of the world from stimuli and to connect it with what we remember, need, and want. Looking at the nervous system from sensations that flow into the brain and actions that are initiated there, it would seem that there's some sort of intelligence in the middle of all the action, working things out. But why does such a seemingly logical organ often have so much difficulty with PI and distinguishing what is real from what is not?

In the next chapter, we'll look at the brain's remarkable power to heal us when we focus on the right perceptions and thoughts — and how we become hopeless neurotics when we allow our fears to usurp our complex mind-body connection.

Mind Over (and Under) Matter

Self-Healing and Self-Sabotage

If you are among the millions of people who suffer from back pain and similar chronic ailments, you may have been dismissed by legions of friends, family, and even colleagues who have repeatedly told you, "It's all in your mind." That helps a lot, doesn't it? In the meantime, the nagging ache in your back is so sharp you can't sit, you can't stand, you can't exercise, you can't sleep. It sure feels as if it's in your body, all right.

According to a 2008 survey, approximately $86 billion is spent each year on over-the-counter medications and surgeries — as well as on innumerable visits to chiropractors, orthopedists, physiotherapists, osteopaths, acupuncturists, yogis, Reiki masters, Swedish masseuses, and a host of other specialists — to treat chronic back and neck pain. I know that many people get significant temporary relief from all these treatments, and I would never knock the fine work of a health practitioner, but are the majority of neck and back sufferers really getting better? Is it possible to rid ourselves of our aches and pains simply by "thinking them away"?

Montel's Story

In the foreword Montel Williams recounted his years battling multiple sclerosis (MS) — a particularly challenging disease to treat, despite the four hundred thousand sufferers in the United States and the two a half million worldwide. Montel is by no means the only celebrity to have been treated for MS; the late comedian Richard Pryor, actress Teri Garr, author Joan Didion, country singer Clay Walker, and actor David Lander (fans of 1970s TV remember him as Squiggy from *Laverne & Shirley*) have all suffered from this malady. But whether or not you are a celeb, the symptoms of MS are quite real, and the paths to a solution can be trying indeed.

I couldn't be happier that Montel's news has been encouraging. Certainly, as he recounted, he has a regimen and has worked diligently at stress reduction (which is so critical to alleviating MS symptoms, not to mention most other ailments, to some degree). He is also fortunate to have a wonderful support network of friends and family. But, as I've observed, the most important aspect of his success in fighting this terrible disease has been his *perception* of it, which enables him to maintain positivity and continue to be proactive in his health management. Montel has high PI when it comes to confronting his illness head-on, setting his priorities straight, taking care of his health matters, being mindful, and most of all, not caving in to *false realities* — doom-and-gloom negative thinking and unnecessary and risky medical treatments.

The belief that being mindful of one's perceptions can help alleviate a wide range of medical issues is not just BS. Countless studies show that the human brain is capable of astounding miracles, and some of the world's top practitioners in a range of fields have blended mindfulness philosophy into their practices, in addition to more traditional medicine and treatments. But how exactly does this mind-body connection work? And how do we develop high PI to improve health and avoid illusions that are detrimental to our health?

Let's Go Crazy and Sexy!

You may have heard the story of Kris Carr, an actress and photographer who discovered on Valentine's Day, 2003, that her liver was riddled with cancerous lesions. She was told there was no cure for this rare type of cancer (I'll shorten the never-ending technical name to EHE) and that at best the treatments would only stave off the inevitable. Though she can't be cured and the disease is inoperable, Kris has lived a joyous, healthful life well over a decade since that fateful day. How has she done it? By being *crazy sexy* in her approach to life, which she details in several books, including the popular *Crazy Sexy Cancer Tips,* and depicts in her self-made documentary by the same name.

Kris is a proponent of a special kind of mindfulness, in which she doesn't allow cancer to get the better of her and refers to it as a catalyst for change. The *sexy* of "crazy sexy cancer" refers to being empowered, getting the most out of every second of life, and refusing to allow the disease to define her. Not only did she reboot her life — switching gears from acting to writing and lecturing on the benefits of a healthy lifestyle — but she has created a "crazy sexy cancer" support group (which she calls a "posse") and formulated her own philosophy of well-being that blends Western medicine and alternative care. As *Scientific American* described, "Carr is among a growing number of people living and thriving with cancer, thanks to medical advances as well as a progressive philosophy in oncology that recognizes past mistakes of overtreatment and welcomes alternative medicine as a partner in the healing process."

According to a recent Johns Hopkins study by Professor Lisa R. Yanek, a positive outlook can help decrease the chances of heart attack among those with a family history of cardiovascular disease. Harvard Medical School has pointed to studies showing that optimism not only helps reduce stress and risk of heart disease but also improves recovery from heart surgery, reduces blood pressure, and prevents further attacks. Substantial evidence reveals that *even smiling and laughing* more each day can make you healthier and prevent disease. Those who smile, laugh, enjoy life, and shrug off problems have high PI, often without even knowing it. Perhaps even more substantiated are the myriad benefits of mindful meditation and how it can alleviate the symptoms of a number of ailments, including back pain, psoriasis, insomnia, and even mental illness. In one famous study by Jon Kabat-Zinn, founder of mindfulness-based stress reduction, or MBSR, psoriasis patients who meditated while receiving UV light therapy healed four times faster than those treated only with the light therapy. What you do with your brain can have a massive impact on disease.

Our perceptions of disease become wired in our brains early in life. If we experienced firsthand the prolonged pain and suffering of a grandparent, parent, sibling, or close friend who passed away due to cancer or another life-threatening disease, those memories can haunt us and lead to intense fear and stress, should we find ourselves similarly afflicted. Strong emotions don't necessarily cause illness, but they are most certainly detrimental once a disease manifests in the body. The worst scenario is when fear and stress foster negativity (low PI), which exacerbates and accelerates the illness and/or its symptoms and lowers immunity, leading to a host of other medical complications.

Dr. Daniel Siegel, author of the groundbreaking book *Mindsight,* describes the harm of negative thoughts: "Bringing these negative thoughts, such as fear, hostility, betrayal, or sadness, to awareness is part of basic health, because those thoughts — what in my field are called unintegrated neural processes — are basically like black holes. They have so much gravity to them that they suck the energy out of life. They influence the health of the mind, its flexibility and fluidity, its sense of joy and gratitude. They impact relationships, leading to rigid ways of behaving or explosive ways of interacting. They also influence the body itself, including the nervous system and the immune system." Forgiving past wrongs done to you is a great way to start becoming free of negativity and improving your PI.

Can meditation and other mindfulness practices cure all disease? No, not quite — but as Grandma used to say about chicken soup, "It couldn't hurt." There is substantial science behind the healing power of relaxation and positivity. At the very least, being positive through an illness makes you more pleasant to be around and will therefore inspire others to be of greater assistance to you in your time of need. Having a strong social network (real people, not Facebook friends) and getting therapy can provide ongoing reinforcement for your positivity to remain intact.

Having a positive outlook can also lead people to effective treatment that mainstream medicine has not embraced. In my own medical practice I see firsthand the role optimism can play in experiencing a successful outcome. Dry eye disease affects millions of women in the United States and is often a chronic condition. More than ten years ago I developed a special eye cream that is 95 percent effective at resolving dry eye symptoms, especially in hard-to-treat cases. Over the course of many years I have lectured to thousands of ophthalmologists and written in ophthalmology publications about my research and treatment of dry eye disease. Because no corporate pharmaceutical company provides this cream (it needs to come from a compounding pharmacy) — coupled with the fact that many eye doctors just aren't comfortable prescribing from a compounding pharmacy for dry eye treatment — many women with dry eyes have suffered for years, even decades, without being effectively treated. The optimistic women who refuse to accept the "dry eye prison sentence" take their future in their own hands, research alternative remedies, and discover my treatment.

My patient actress-turned-entrepreneur Victoria Principal is a good example. She experienced dry eyes for years, and no prior treatments resolved the condition. Victoria, who has high PI, remained hopeful that there was an effective treatment beyond what her previous eye doctors knew. That positive mind-set drove her to do her own research that eventually led to me and a successful treatment after years of failed conventional treatments. Having this mind-set can also lead you to find treatments for conditions that conventional medicine thinks do not exist. An example comes from another patient of mine with high PI, Jamal Crawford, an NBA basketball player, who had brown pigmentation or "freckles" on the whites of his eyes for years. It was frustrating for him, since he saw it in photos and recordings when he would replay the games. Doctors typically tell patients with this ailment, "Sorry, there is no treatment. You just have to live with it." Jamal researched and found out about what I do for these brown "eye freckles," and I removed them.

Maintaining a positive, optimistic attitude reflects high PI and can lead you to find breakthrough medical treatments for what otherwise might seem like hopeless scenarios.

You can influence your own PI when it comes to health issues. Positivity breeds positivity, as it did with Montel Williams and Kris Carr. If you think about it, negativity and woe-is-me behavior about aches and pains might have a lingering impact on those you leave behind — a spouse, a child, a grandchild — which can cause them unnecessary fear and suffering. While it is important to get yourself checked out by medical professionals for symptoms that may or may not turn out to be of concern, continually complaining about them is generally not productive. It's far more healthy and effective to once in a while vent to a tolerant confidant or therapist to get it out of your system and release any frustration about your maladies all at once. On the other hand, too much complaining to friends, family, and coworkers, and you could easily be perceived as an all-around kvetch. Before you go down the long and winding road, ask yourself: *Is this how I would like to be remembered?*

Our Bodies, Our Minds: The Grand Illusion

What is an illusion anyway? Illusions are distortions of our perceptions that can appear in myriad forms but that primarily affect our senses. Illusions commonly occur with regard to visual perception. The typical Hollywood film representation of this is the hero, dying of thirst while lost in the middle of a barren desert, suddenly imagining a mirage. We could call this an outright hallucination.

Illusions happen all the time, since our brains can trick any of our senses, convincing us that the impossible is reality. There are numerous case studies of amputees who continue to experience pain in the area of the missing limb long after the limb is gone. Some scientists attribute this phantom pain to nerve endings at the end of the remaining wounded limb; others believe that a memory of the limb lingers in the brain. I propose a middle ground. Our memory of the limb and prior perceptions of what pain feels like may have kicked in to make these sensations seem real. In cases where nerve endings are transmitting the pain, the brain has no other method of interpreting the sensation, so it diverts back to a familiar memory of the limb being present.

Hypochondriacs in particular reside in a grand illusion that warps and lowers their PI. That's not to say their perception of serious pain or illness isn't real to them. If it weren't, these one in twenty people who needlessly venture to and from medical visits wouldn't be wasting so much of their time loitering in waiting rooms, throwing out money on copayments to doctors' offices, and expending emotional energy on chronic worrying.

In order to be officially diagnosed with *hypochondriasis* (also known as Illness Anxiety Disorder, or IAD) according to the *Diagnostic and Statistical Manual of Mental Disorders* (*DSM*), one must be convinced of the ailment for at least six months, despite medical evidence proving otherwise. Hypochondriacs emphatically believe in their conditions and truly suffer; many of them fall into deep depressions. Why have their perceptions gone so terribly awry?

FAIRY TALES AND FALSE MALADIES

If you're a hypochondriac, you can feel somewhat better knowing that at least you're in good company with some major creative talents, including Tennessee Williams, Marcel Proust, and Andy Warhol. Hans Christian Andersen, the nineteenth-century Danish children's fairy-tale writer, was also a noted hypochondriac. Not only was he convinced that a tiny harmless mark above his eye would spread over his face, but he had a lifelong phobia of being buried alive and carried around a note that said, "I only seem dead." Andersen left behind a timeless legacy of children's literature. Among these classics is "The Emperor's New Clothes," a famous story about mass influence on PI: if the king believes his invisible suit of clothes to be real, then his people also purport to see it as well. (See chapter 12 for more on the concept of following the herd.)

Hypochondriacs accustomed to hearing the cliché "It's all in your head" are suffering from a very real medical issue: anxiety disorder. At the root of it is low PI; the mind becomes stuck in a loop of signals warning that something is terribly wrong (when, at least at first, there isn't). The illusion becomes so convincing that the mind translates the signal into perceived pain that, by all accounts, feels quite real. Psychologists speculate that hypochondria can stem from a number of potential causes, including an anxious, smothering parent who also happens to be a hypochondriac (neuroses and fears passed down); a reaction to a family death or illness; a past trauma; or physical or emotional abuse received as a child.

Some experts now theorize that the barrage of symptoms and diagnoses available through search engines such as Google and prevalent in online surveys can trigger hypochondria, simply owing to the power of suggestion. One can't help noticing the irony of our modern-day glut of free, 24/7 medical information causing so much unnecessary anxiety — a real medical condition — that never would have existed before the Internet came into being. Technology cuts both ways.

Our PI is in constant overdrive as it sifts through the sheer volume of health scares posted every day on medical and news websites and in blogs. Our social media networks allow us to conveniently share articles among our network of friends and family, spreading the viral information, whether or not it's valid. (A rash of recent viral posts inaccurately said the Centers for Disease Control advised mothers to stop breastfeeding in order to increase the effectiveness of vaccines.) If we see a health article about a disease on Facebook that we also received via email from our sister, cousin, best friend, and work colleague, our perceptions may become distorted into thinking that we might actually have something seriously wrong — which can lead to everything from insomnia (due to chronic worrying about the ailment and then more googling) to depression. The Internet is a world brimming with endless data — not always accurate — that has the potential to turn us all into hypochondriacs, if we don't hone our PI to help us judge what is real and what is not. With insight and treatment, those with hypochondriasis can improve their Perceptual Intelligence to better recognize their health illusions.

As for those who fear that there are little green men with laser guns under the bed — well, that is an entirely separate type of illusion that we'll address in the next chapter.

3 What You See Is Not What You Get

Mind Tricks and Illusions

Harry Houdini, the most famous illusionist and escape artist of the early twentieth century — and perhaps of all time — was quoted as saying, "What the eyes see and the ears hear, the mind believes."

Human beings love mysteries. We are fascinated by illusion and magic, which is why we don't mind putting our reality-versus-fantasy meters temporarily on hold, willingly lowering our Perceptual Intelligence, and allowing our imaginations to be dazzled by entertainment. This is why people flock to see such talented entertainers as Penn and Teller, David Blaine, and David Copperfield. Separate from magic exists a genre of illusionism known as *mentalism* in which performers, such as Derren Brown and Max Maven, employ clever trickery to make it seem as if they can read or control minds. (I am convinced that my eleven-year-old twin daughters are working on this type of sorcery.)

Some skilled magicians and mentalists readily admit in front of their audiences that there is no psychic phenomenon behind what they do. Penn and Teller, in particular, famously disclose how the tricks are done yet delight audiences anyway because their sleight of hand and other professional methods require so much practice and training that theatergoers still can't keep up with their magical mastery.

Other individuals are reported to have accomplished genuine feats of psychic illusion. In the late 1970s the Defense Intelligence Agency created a special unit known as the StarGate Project to investigate psychic phenomena and determine if certain psychic ability could be measured and mined for military purposes. Sounds like the *X-Files*, right? At the time the government believed the project had enough potential that it was worth investing our taxpayer money in testing such famed individuals as Uri Geller and Ingo Swann for their abilities, notably with regard to *remote viewing* — being able to see objects in other places (in the case of Swann, unseen aspects of other planets such as Jupiter) without ever having been there. To this day, many of their feats haven't been fully explained by science.

Then, of course, there are the skeptics — notably, James Randi. An accomplished magician himself, the "Amazing Randi" has spent a lifetime debunking psychic phenomena. He once remarked, "Uri Geller may have psychic abilities by which he can bend spoons; if so, he appears to have been doing it the hard way." Personally, I think Randi is just jealous. Margaret Thatcher once said of her detractors, "If my critics saw me walking over the Thames River they would say it was because I couldn't swim."

It's perfectly fine to take magic and illusion at face value and not question the art forms; it's all entertainment, pure and simple. Audience members may say, "I loved that trick even though I don't have a clue how he performed it." The trick is admitting that *it's a trick*. The tipping point to low PI occurs when people are convinced that what they witnessed on stage is 100 percent real, no matter how far-fetched.

On the other end of the spectrum are the hardened skeptics whose thinking is so concrete that they cannot conceive of anything they can't personally explain by logic and sight, hearing, feel, and touch. We all have varying degrees of PI when it comes to the world of illusion, and most of us reside somewhere in between these extreme perspectives.

For the majority of us, however, it is often difficult to accept the perspectives of those who keep hold of their illusions even when they are proven to have no basis in reality. Since our "realistic" PI doesn't register and accept these illusions, we pooh-pooh the believers — no matter how otherwise intelligent they might be — and label them as gullible, or we use less diplomatic terms. Why do their "PI receptors" readily enable them to accept their illusions as real?

To a large extent, the believers *want to believe.* They enjoy the illusions and, consciously or subconsciously, wish them to be true to such an extent that they allow their PI to accept them as authentic. They want to believe in magic and illusion because it's fun to imagine people who have "special powers" and can do remarkable things. (Witness the never-ending blizzard of superhero films and TV programs, from *Wonder Woman* to *The Avengers* to *Superman* to *Batman* to *The Defenders* to *X-Men.*)

For better or worse, there's no getting around the fact that we're irrevocably reliant on the three-pound blob of sensitive biological tissue protected within our skulls to help us separate reality from illusion. As this chapter unfolds, you'll learn that our minds can perpetuate illusions that seem as real as anything else in front of us.

Dark Lucid Dreams Are Made of These

An acquaintance of mine once told me that when he was a five-year-old living in a two-story house in Queens, he saw an alien hovering outside his upstairs bedroom window. More than four decades later, he makes light of it and logically knows that an alien never appeared outside his window; it was all an illusion. He doesn't in any way purport that aliens exist, and yet — to this very day — he vividly recalls the green face and antennae pressed in close against the glass, the bug-eyes staring right at him. His memory of the illusion remains strikingly real to him: Why has it stayed so fresh in his mind after all these years?

Millions of people, many of whom are certain they were

abducted, have claimed to see aliens. Outside conspiracy theories, weather balloons mistaken for UFOs, alleged cover-ups of bizarre events (such as in Roswell, New Mexico), "firsthand" accounts, and so forth, no solid evidence supports the belief that we regularly receive alien visitations. To the question of whether aliens exist, Carl Sagan once replied, "I give the standard arguments — there are a lot of places out there, the molecules of life are everywhere, I use the word billions, and so on. Then I say it would be astonishing to me if there weren't extraterrestrial intelligence, but of course there is as yet no compelling evidence for it."

Why then do so many believe that the "truth is out there," as Fox Mulder from *The X-Files* put it? Until proven otherwise, scientists will continue to purport that alien experiences and many other phenomena are vivid illusions conjured up by our imaginations while we sleep. *But how can these illusions seem so real?*, you may ask. More likely than not, individuals who experience them are suffering from a common disorder known as *sleep paralysis*, which reportedly affects four in ten people.

People with symptoms of sleep paralysis feel as if they are wide-awake when they are actually either in a state of falling asleep or in the process of waking up. They are between conscious and unconscious states and are tricked by their brain into thinking that everything occurring in these moments is transpiring in reality. This is when their "dark lucid dreams" (the extreme version of "lucid dreams," a term coined by Frederik van Eeden in 1911) take charge of their brains and senses, allowing their imaginations to run amok. Floating between the worlds of the conscious and subconscious, people with sleep paralysis can't move (though they may try to) when they are in this state. They are frozen and helpless as aliens or strangers appear by their bedside. Or they may feel as if they are being struck by a blinding light, transported aboard a ship, laid down on a metal slab, and experimented on by little green beings. When they awake (for real) they are paralyzed by fear, often reliving the residual physical pain and other terrors inflicted on them while they were buried inside the dark lucid dream.

But why do alien confrontations, specifically, come up so often for people? When we are trapped in sleep paralysis, our minds don't know what to make of these illusions and are befuddled as they search for meaning. Our senses are lit up to such a degree that the signals they send to our brain are interpreted as something familiar in our imaginations: *aliens*. Somewhere deep in our minds is the implanted memory of an alien from a science fiction film, TV show, book, or even a drawing. While the trigger of the image might be rooted deep in our psychology — such as a feeling of powerlessness caused by a hidden childhood trauma — the translation of this dreamy illusion formulates in our minds, and we believe these aliens to be real.

Daily we are surrounded by images that get tucked away in our memory banks. As it happens, the acquaintance I mentioned earlier in this passage admits that he frequently watched the original *Star Trek* TV show and other science fiction programs with his dad, starting at an early age. Isn't it possible that the vision of an alien jumped to the foreground of a dark lucid dream because his young, impressionable mind was still processing a colorful image he'd seen on TV?

When our brains can't identify illusions — such as during a dark lucid dream — our mind convinces us that they happened because we can't conceive how such dramatic emotions and sensations could be fake when they *feel* so real. I have had firsthand experience. The night before working on this chapter I experienced an episode of sleep paralysis. I was staying alone in a hotel in Chattanooga, Tennessee, for a team rowing competition. While I was sleeping, I thought I had awoken to a man standing beside my bed who held my arm and shoulder, pinning me to the bed. I couldn't move any part of my body. I also couldn't yell for help. Finally, I woke up to realize I had experienced a dark lucid dream, a.k.a. sleep paralysis.

During an episode of sleep paralysis, images appear in our

heads as if we had seen them with our own eyes. Long after experiencing these events we can still describe every powerful sensation, right down to the pain of an alien jabbing a needle into our arms and the freakish otherworldly smell of the experimentation room aboard the alien spacecraft. It's scary, to be sure, but in reality it's just our minds messing with our lowered PI.

I can't claim that all alien sightings fall into the realm of dark lucid dreams, however, or that low PI is behind every experience. There are also documented incidents in which mysterious objects have been spotted by *groups of people* (a few during the day). These events are photographed and videotaped and, yes, at least some (or many) could have been weather balloons or could have involved doctored images (easy enough to do these days) — but there are also incidents that continue to defy explanation.

Ana Zamalloa, an educated and respected Peruvian tour guide with twenty-three years of experience, recently led my wife and me around Peru for a week, during which time we became closely acquainted. While we were at Machu Picchu, Ana explained that this geography was a sacred, mystical place for the Incas. She knew of many people, including a friend of hers, who reported seeing spirits or ghosts there at night. (It's actually no longer open in the evening.)

Ana shared a story from 1999 about a client, Jerry Wills, an energy healer from Arizona who was on vacation in Peru, with Ana as his guide. Jerry explained to Ana that he had been deposited as a baby on Earth from another planet. She laughed and poked fun at him. He wasn't offended in the least, but in a casual voice said, "Tomorrow my friends will come."

The next morning she took Jerry to tour Huayna Picchu Mountain. Jerry called to Ana and pointed across the canyon to the mountain to the right of Machu Picchu. Through her binoculars, she witnessed an object approximately twelve to fifteen feet in diameter appearing just under the peak of the mountain, a location inaccessible to climbers. It was a large, gray, metallic-looking sphere with black on top, and it looked nothing like a weather balloon (or a balloon of any kind). The orb moved back and forth. Ana blurted out, "Oh my god!" and studied it along with the group of tourists for about fifteen minutes before resuming the tour. She explained to me that she had never seen anything like that since. She was not dreaming or hallucinating, since the amazed tourists saw the orb as clearly as she did; in fact, one member of the group videotaped the sighting and played it back. Neither did Ana feel paralysis or fear, which would be typical symptoms of the dark lucid dream of sleep paralysis.

What happened here? Is this a case of low PI? I would say it's not, and here's why: not only was Ana credible as a person and in her storytelling, but she also didn't attribute what she saw to any specific extraterrestrial or supernatural phenomenon, nor did she hype it in any way. She simply *reported what she had seen*, which leads me to accept that she did, in fact, see something strange out there. Until science proves something specific, it's perfectly acceptable to leave the question open or just to speculate. The acknowledgment of an unexplained mystery — along with a healthy dose of scientific skepticism — is what sparks our imaginations and fills our world with continuous wonder.

Paint It Black: Is Art an Illusion or a Lie?

Pablo Picasso once said, "We all know that Art is not truth. Art is a lie that makes us realize truth, at least the truth that is given us to understand. The artist must know the manner whereby to convince others of the truthfulness of his lies."

What was Picasso talking about? Where's the "lie" in a still life of a vase of colorful flowers? And if all art is a lie, then why should we have museums, exhibitions, and exhibits? Why do we need artists, including Picasso? While there is no shortage of people who enjoy spinning a good fib, who wants to be lied to?

To explain why we have this need, we must look to the brain, which helps us make sense of the mosaic of lines, colors, patterns, and images that we think of as art. Long before human beings had any knowledge of neuroscience, artists created appealing, provocative illusions of people, places, and things that weren't really there but seemed very real. Artists bend our Perceptual Intelligence, constantly challenging us and demanding that we pay attention and interpret their work.

For centuries, artists have used specific colors to convey the impression of depth in paintings, an effect known as *chromostereopsis*. (I dare you to say that ten times fast.) Red, for example, will appear to advance and blue to recede, which is why blue and other cool hues are used to convey distant images.

Color and luminance are fodder for artists. The retinas of most human eyes contain three kinds of cones: red, blue, and green. You know what color you're looking at because your brain compares the activities in two or three cones. A different phenomenon, called *luminance* — the amount of light energy emitted or reflected from an object in a specific direction — adds the activities from the cones together as a measure of how much light appears to be passing through a given area.

To make something look three-dimensional and lifelike, artists add elements, such as lightness and shadows, that wouldn't be present in real life but that tap into our hardwired expectations of "what should be," thereby fooling our brains.

When it comes to art, our brains readily identify faces as well — even when we're viewing images constructed from colored lines, color patches, or disparate images (think of a Chuck Close painting). Researchers have found that the amygdala, the part of the brain involved in emotions and the flight-or-fight response, responds more to blurry photos of faces expressing fear than to unaltered or sharply detailed images of these faces. At the same time, the part of our brain that recognizes faces is less engaged when the face is blurry, signifying perhaps that we're more emotionally engaged when the detail-oriented part of our visual system is distracted, such as with Impressionist works, in which faces are often unrealistically colorful or patchy.

PUTTING IT TOGETHER:
THEATER, MUSIC, ART, AND ILLUSION

Lest we think that painters are the only ones capable of toying with our PI, let's take a look at one notable contemporary work and how it interweaves theater, music, and art to explore the creative process and even question reality. The Pulitzer Prize–winning musical *Sunday in the Park with George*, with music and lyrics by Stephen Sondheim and book by James Lapine, showcases how art is created from a variety of perspectives. In the first act, we view a fictionalized version of nineteenth-century Impressionist painter Georges-Pierre Seurat as he creates his masterpiece, *A Sunday Afternoon on the Island of La Grande Jatte*, directly on stage. (First he makes sketches in the park, and later he adapts them to the larger-than-life painting in his studio.) While he does so, we see him conjure and organize the people in the painting; control the environment (at one point drawing and then erasing a tree, which humorously disappears and thus disturbs the reality of one of the characters); sing through the characters (including a dog!); and fall in love with (and squander the affections of) his model, Dot. The music, lyrics, and Seurat's physicality while painting fuse to reflect and translate the artist's revolutionary pointillist style of using tiny dots (hence his lover's symbolic name).

The real-life artist Seurat manipulated colors and light through dotted brushstrokes — for example, putting blue and yellow next to each other to create the illusion of green — and we are fed an opportunity to simultaneously grasp this miraculous artistic concept in motion and get visually fooled by the illusion of the blended colors (reduced PI). When the true-to-life painting coalesces with the actors frozen in position blending into it, we feel as if we have witnessed art being created right before our eyes on stage. At its best, theater engages, stimulates, and manipulates our PI and deepens the relationship between actors and the audience; at the same time, it creates the illusion of another reality in real time, and we willingly go along for the joyous ride.

One of the paintings most etched in our minds is *The Scream*, Edvard Munch's 1893 Expressionist masterpiece. When we visualize the work, we automatically feel the torment of the figure in the foreground and may even recall the swirls of color receding behind him (or her — the gender is unclear). Munch has created a powerful illusion; he has translated the sound of a scream into something that can be *visualized* through the waves swirling on the other side of the railing. The scream is so potent it even converts the person portrayed into an almost liquid substance. The painting functions on so many levels not only because of the emotional reaction it elicits in the viewer but also because the event is believed to have been autobiographical. The conception of the work derives from Munch's explosive reaction after two people, visible in the background, had left him. Whether or not he was aware of it at the time, Munch effectively distorted our minds into believing that sound emanates from his painting and that it physically disrupts the environment and the human forms to the point of changing their shapes — low PI on a granular level on the part of the viewers.

Another Expressionist painter, Vincent van Gogh, suffered enormously from his vivid perceptions, which he converted into distorted visions and dreams painted on canvas. His Perceptual Intelligence was way off-kilter; a haystack was not just a haystack to van Gogh, who interpreted it as a living, breathing object. It is entirely possible that this is exactly how he saw the haystack in his head as he studied and painted it.

There is hardly a painter whose life and work (and the relationship between the two) have been more analyzed than van Gogh's. Whether he had undiagnosed temporal lobe epilepsy, lead poisoning, bipolar disorder, depression, or schizophrenia, as various people have suggested, we will never know for certain. Whatever the cause of his altered PI, we may deduce that his brain constantly played tricks on him; he believed in the illusions strongly enough that he was driven to capture them in his art.

Blinded by Science: Illusions Can Be Proved to Be, Well, Illusions

We've spent this chapter exploring how our PI can be hijacked by various illusions that manifest in magic, perceived illness, dreams, and art. What does the grounded, hard-core world of science have to say about *why* we maintain the world of illusion? In his work *Phaedo*, Plato wrote: "Is there any certainty of human sight and hearing or is it true...that we neither hear nor see anything accurately?" He later adds, "Observation by means of the eyes and ears and all the other senses is entirely deceptive."

As a medical professional, I would never question certain scientific truths: how the eye detects light, how sound travels through the auditory canal, how our skin feels touch, and so forth. And yet scientists revel in challenging existing science because so much remains unexplained, and uncovering more data could lead to better treatments. We should never accept the status quo but should instead always seek to improve our understanding of science. If I hadn't threatened the beliefs of many of my fellow eye specialists and surgeons by creating a noninvasive treatment called Holcomb C3-R® (named after US Olympic gold medal bobsled driver Steven Holcomb) to treat Keratoconus, many thousands of patients would have otherwise undergone invasive and painful cornea transplants.*

Illusions surround us all the time, and how developed our Perceptual Intelligence is drives whether we accept them as real or not. Late at night when you're alone in your pitch-black living room, does the coat on the hook start to take shape as a human being? Your reaction, of course, is to turn on the light just to be sure your PI isn't being fooled, even though logically you know it must be a coat you are seeing. Let's take it a step further: If there is a power outage in the middle of the night and you don't have any candles or flashlights, might that coat start to look like a prowler with a weapon? Let's say that this takes place on Halloween, you've had way too much candy corn, and there is a howling thunderstorm outside. Your PI suddenly has undergone a hostile takeover and sees that coat as a horrific ghost or monster.

Illusions can be perceived on different levels and can vary from one person to the next, depending on brain anatomy. In one study, a group of people was asked to look at a Ponzo illustration, a geometrical optical illusion that shows a receding train track with two horizontal yellow bars. Mario Ponzo, the Italian psychologist who developed this type of illustration, suggested that the human mind judges an object's size based on its background. Researchers used high-resolution fMRI (functional magnetic resonance imagining) technology to scan the brains of the subjects while they were viewing the illustration. The size of the illusion was established by asking them how much larger the lower bar had to be to make it look the same size as the upper one. While all the study subjects perceived one bar (the upper one) as larger, the effect's magnitude differed substantially across individuals. These differences were found on the surface area of the primary visual

---

* You can find my TEDx Talk on my fight against the medical establishment for sight for Keratoconus patients on YouTube: www.youtube.com /watch?v=7RUN9wK0uPA. Or save some typing by searching YouTube for "boxer wachler tedx."

cortex, located at the back of the brain. The researchers found that the smaller a person's primary visual cortex, the more powerfully he or she experienced the illusion. While the primary visual cortex influences the degree of the illusion, it is the individual's PI that is affected in the process as he or she determines if the illusion is real.

In my freshman Psychology 101 class at UCLA, I learned of a study involving college students wearing prism glasses that flipped the world upside down and profoundly illustrated the malleability of their perceptions. Perceptions are like muscles being trained, especially when it comes to avoiding confusion. The students had to wear the glasses 100 percent of their waking hours, for days at a time. When the specialized prism glasses were worn, they wreaked havoc on the subjects' vision and perceptions. Can you imagine trying to drive when everything looks upside down? After a week of wearing the glasses, however, the subjects found that their vision had adapted and that the world looked normal again. When they removed the glasses, they were disoriented again for a short while until their vision readjusted. This is a powerful scientific demonstration of how the brain acts like a muscle and has the capacity to correct a known illusion — even though it is creating a brand-new illusion in order to do so.

At the conclusion of a speech at the Being Human 2012 Conference, neuroscientist and artist Beau Lotto stated: "Either there are no illusions or everything is an illusion. And given that we are pretty much all delusional, you might as well choose your delusion." This statement may serve for the type of illusions described in this chapter, but in the next, let us see how well it holds up when our Perceptual Intelligence goes up against the ultimate illusion: death itself.

4 Out of Body or Under the Ground

PI and Experiences of Death

"I don't believe in an afterlife, although I am bringing a change of underwear," quipped the title character in Woody's Allen's humorous short story "Conversations with Helmholtz."

For all our scientific study and research, we know nothing whatsoever about what comes after existence — most specifically, whether there *is* anything. To this day, no one has scientifically proven anything concrete with regard to what lies beyond. As I mentioned in chapter 1, we have only scratched the surface in our understanding of all the intricacies of the brain. Multiply our knowledge gap by about a zillion, and we'll be somewhere in the domain of what we know about death and the hereafter.

With such a lack of proof, how can we be 100 percent positive of what is reality and what is fantasy and therefore know on which end of the PI spectrum we reside? Many scientists and philosophers — including men and women of strong faith — won't dismiss the possibility of an afterlife or a soul. The believers don't need the claimants who died and came back to provide passports stamped "heaven's gate." Among those medical practitioners who believe and have written extensively on the subject are Dr. Raymond Moody, author of *Life after Life*; Dr. Mario Beauregard, coauthor of *The Spiritual Brain*; Dr. Mary Neal, author of *To Heaven and Back*; and Dr. Jeffrey Long, author of

Evidence of the Afterlife. Dr. Long is also founder of NDERF (the Near Death Experience Research Foundation), an organization that has collected more than four thousand near-death experience (NDE) stories and shared them on its website. Can all these medical minds be right, even though their claims aren't rooted in science as we know it? Is there some scientist conspiracy at work?

Does my belief in a "soul," as a medical practitioner, validate the authenticity of all afterlife stories? Yes and no. The soul may indeed exist and there might, in fact, be an afterlife. (In fact, as you'll discover as this chapter unfolds, I have a personal reason to believe in out-of-body experiences.) The fact remains, though, that we lack any proof of an afterlife. What we do know is that our innate PI registers our ability to distinguish between reality and fantasy, allowing us to weigh in on and visualize concepts we can't possibly comprehend about death and helping to provide meaning and comfort as an alternative to nothingness. If our PI is low in this regard, we fill in the blanks and are presented with what our minds perceive as "sneak peeks" into what lies beyond. It is those areas of investigation that continue to intrigue, perplex, and haunt us: near-death and out-of-body experiences.

The Light at the End of the Tunnel

To find statistics on NDEs, we must go as far back as a 1992 Gallup poll in which 13 million Americans reported they had had sneak peeks into the world to come. The numbers are significantly higher today, especially given the advances made in science and all the doctors saving lives in ways previously thought impossible. Given the large number of people who have had NDEs, it is likely you know someone who had one, or maybe you have experienced one yourself. All the fanfare about these afterlife accounts has inspired new ideas about human consciousness, and even acclaimed neuroscientist and atheist Sam Harris has speculated about the idea of consciousness outside the brain. Though there are myriad unique stories, the most commonly repeated scenarios reported include:

- Seeing a blinding light, often after entering a tunnel
- Watching a life playback, as in "my life flashed before my eyes"
- Receiving visits from long-deceased relatives: "seeing Grandpa"
- Entering heaven, taking a journey through a beautiful place, such as a garden
- Meeting God himself, typically a white-haired figure in the tradition of Michelangelo's painting on the ceiling of the Sistine Chapel
- Having an out-of-body experience (which we'll cover later in this chapter)

Some fairly well-known accounts of NDEs have drawn more than just a bit of popular attention. There is the story of four-year-old Colton Burpo, who was dead for some three minutes during an emergency appendectomy. When he miraculously came back to life, he was able to recount his parents' activities during his operation; a hug with his unnamed baby sister who had died in the womb and he hadn't known existed; and an elaborate description of his visit to heaven that reflects Bible passages he couldn't have heard or read at that age. Colton's father wrote about his son's experiences in his book *Heaven Is for Real*, which has sold millions of copies and been made into a film.

More startling is the tale told by Harvard-trained neurosurgeon Dr. Eben Alexander in his *New York Times* bestseller, *Proof of Heaven*. After being in a seven-day coma suffered as a result of E. coli attacking his brain, Dr. Alexander awoke just as the doctors were ready to stop treatment. Not only did he make a miraculous full recovery, but he claimed that during his absence he had been touring heaven with the divine source. Imagine the outcry from the medical community that someone as educated and reputable as Dr. Alexander would not only convey his NDE as "fact" but would also publish his story and feel compelled to go on a crusade preaching NDE gospel.

There are many other distinguished, credible people who, like Dr. Alexander, feel the need to share their stories and convince others of their authenticity. Some came back with a renewed sense of purpose; others with long lists of things that needed to be accomplished. Whatever the case may be, I do not doubt for a minute that they are thoroughly convinced their NDEs were real and that these experiences subsequently led them to follow their controversial paths. While I am a believer in the concept of NDEs and out-of-body experiences, I also propose that logic and science can help us better understand these experiences. I am convinced that the unexplainable and science will someday meet, just as we are able to accept the ideas of a God and a soul existing alongside the fact of evolution. Meanwhile, there is some science we do know, particularly with regard to how we connect the subjective nature of death with where our Perceptual Intelligence stands during our most fragile, teetering moments.

I'm Not Dead Yet

What happens to you when you exit the world of the living and reemerge to tell the world about it? Before we hasten through the valley of death and address what is happening during NDEs and its connection to PI, we must first address one critical question: At what point is a person truly dead?

When a patient stops breathing and the heartbeat flatlines, the onus is on the doctor to pronounce the person clinically and officially dead. But this "time of death recording," as it becomes legally documented, is subjective; another doctor might have called it seconds or minutes before or after (or not at all). But are the heart and lungs the be-all and end-all of determining life? It's a well-established medical fact that brain activity continues for up to thirty seconds after all blood flow in the body stops: If those synapses are still jumping, is the person considered alive?

In 2013 the University of Michigan conducted a study with rats in which scientists found that highly synchronized brain activity occurs thirty seconds after the animal's heart stopped beating. During that time, the rats exhibited signs indicative of consciousness and visual activation. Two questions arise from this result: Are the animals alive? Or are they perceiving something from beyond our normal spectrum? Given neuroscience's recent success at establishing the neural correlates of consciousness, many would argue that this last electrical impulse is a mere random firing of neurons — one last bow, as it were, before the curtain drops.

In a different study of rats conducted in 2015, also by the University of Michigan, researchers drew a startling conclusion: When faced with cardiac arrest, the brain sometimes sends a signal for the heart to shut down in order to *accelerate death*. The new theory proposed is that cardiac arrest patients can potentially be saved by surgeons *blocking* this brain activity as they repair the heart to avoid the shutdown signal.

The above suggests that life-forms have their own hidden self-destruct systems — but why? One can speculate that the brain has a do-or-die exit strategy in the form of a machinelike "off" switch. It sees no other way out and wants to protect us from such unresolvable pain, so it quickly throws in the towel as a form of programmed self-protective mercy. It's almost as if once the switch is flipped, the brain wants the heart to stay off.

This brings us back full circle to NDEs. One-fifth of cardiac arrest survivors report seeing visions or having other perceptions during clinical death. If the brain is in fact sending signals to the body to self-destruct during a perceived fatal attack, perhaps it is reasonable to theorize that it is also attempting one of two other protective tactics: interpreting what it can't process or accept (i.e., death) or recognizing what is occurring and providing comforting imagery to numb or distract the dying individual. In the latter scenario, the brain lulls the body into letting nature take its course and finding peace, the equivalent of a movie short that reveals whether we have low or high PI.

There are other potential explanations for these final electrical brain surges. Some scientists point to there being too much carbon dioxide in the blood. The results of a recent University of Kentucky study suggest that NDEs are really an instance of rapid eye movement (REM) intrusion. In that disorder, a person's mind can wake up before her body, causing hallucinations to occur. Cardiac arrest could trigger an REM intrusion in the brain stem, the region that controls the body's most basic functions and that can operate independently from the newly deceased higher brain. If we have low PI, the resulting NDE is the short film (or dream) that our minds interpret as something otherworldly.

Perceptually in Motion

Time will tell if the 2015 study of dying rats conducted by the University of Michigan cited earlier will lead to broader conclusions related to NDEs. Will disconnecting the brain from the heart prevent death caused by cardiac arrest (i.e., stop the "off" switch from being flipped)? Does blocked brain activity for humans during these episodes mean that people who die and come back will not experience NDEs?

The skeptics would answer yes to both questions. NDEs have been known to occur among people of all ages, religions, races, creeds, and colors, and yet it's exceedingly rare to find an instance in which the individual sees a vision outside his or her background, belief, and frame of reference. That admittedly is one strike against NDEs as fact. The late Elisabeth Kübler-Ross, psychiatrist and author of the seminal works *On Death and Dying* and *On Grief and Grieving*, also published a collection of essays based on her encounters with twenty thousand people who had had NDEs. She acknowledged culture's effect on the perception that consciousness continues after death. "I never encountered a

Protestant child who saw the Virgin Mary in his last minutes, yet she [the Virgin Mary] was perceived by many Catholic children."

How can all these different religious allusions be so radically different? Does this suggest that there are separate, customized versions of heaven for every potential affiliation (and, if so, can you preorder it on Amazon)? Or do people not wish to accept or divulge a vision on the other side that does not match their belief systems in our real world? Our minds introduce us not only to the *worldview* we choose to accept (high PI) but also to the *otherworldly views* we choose to accept (low PI). Until science yields more information, we can't entirely refute the idea of religions coexisting in a hereafter or the fact that the messages received through NDEs are designed in our minds in the best way possible for us to embrace them. In these instances, having either low PI (being a believer) or high PI (being a skeptic) is not necessarily a bad thing — unless science irrefutably disproves the realness of these near-death experiences.

Of course, it would be difficult to assess the damage that would be done if everyone were to blindly accept a blissful heaven on the other side: Would there be a risk of significant increases in suicide? Frankly, if your life sucked and science had proven that heaven was like that depicted in the Michelangelo painting, suicide might seem like a pretty good option. My advice: don't do it. Life is too precious.

With a few notable exceptions, people's reports of their NDEs are also relatively free of divine wrath. However, it may be that the people reporting NDEs are generally benevolent people with good standing in society. It would be interesting to hear an NDE account from a murderer or rapist, if any such criminal had an NDE to report. Would the afterlife look less rapturous to *that* person?

As for believers in NDEs, like myself, there is another way to regard what might occur if the brain were to be blocked, preventing an NDE. Although a soul might be making its excursion to the other side during this time, it seems our minds wouldn't be in a position to play back the experience because the part of the mind responsible for the imagery wouldn't be functioning to make sense of it for us. Our minds require brain signals to explain the unexplainable and/or give us a peaceful, if not revelatory, film with familiar images to take our minds off the ultimate curtain closing. When we are "between worlds" our minds are still functioning in unseen ways and tapping into images, ideas, characters, and stories that provide us with comfort and help us explain sensations and perceptions we do not understand (or do not psychologically wish to do so).

Consciousness while Unconscious

The hypotheses outlined so far in this chapter still don't account for those people who report seeing or knowing things during their NDEs that seem impossible, such as young Colton Burpo stating that he spent time on the other side with his sister, who died in the womb and he couldn't possibly have known existed. There are also unexplained cases of people who discover during an NDE that they were adopted or who cross paths with relatives they never met and describe their personal characteristics to a tee. What is the role of our PI in these remarkable circumstances?

The secret lies in our consciousness — a hotly debated issue among scientists. Recent breakthroughs in neuroscience suggest that consciousness lies well within the realm of scientific inquiry. Even when we are unconscious, our brains are working to interpret disorienting experiences. No one would disagree that the brain is still active in these moments. In that gray area we connect the dots within the deep recesses of our minds until we find something familiar and coherent. Whether the sounds or pictures in our minds are based on buried memories or formulated by our imaginations, they are dependent on the familiar — what we know or at least have been conditioned to accept as true. A child like Colton Burpo, for example, will view his NDE through the prism of his religious upbringing and Sunday school experiences.

Whether Colton overheard his family talking about his lost baby sister (while he was conscious or unconscious) or intuitively read something between the lines, we will never know for certain.

We can also speculate that the Burpo family was predisposed to accepting miracles, given the remarkable nature of their son's return. Their PI was perhaps inclined to welcome the idea of their son seeing their unnamed daughter because they likely wanted to believe she had moved on to a happier place and had grown up in the beyond. Their utmost wish was to be united with their baby girl on the other side and someday even bestow a name on her. Given this understandable desire, it's not out of the question for family members to have unknowingly planted subtle hints in front of their son and reinforced aspects of a biblical heaven.

None of this is intended to "expose" the Burpo family — or anyone who has reported an NDE, for that matter. I don't deny Colton's belief in his experience or his family's. Rather, I am suggesting that while we are unconscious, there still exists a degree of consciousness allowing us to see, understand, and piece together fragments informed and influenced by familiar images, experiences, and backgrounds. Our minds still need to process and convey what it is reading from the "other side" in ways that make sense to us. It might be said that those people close to NDE survivors have low PIs in the sense that they are willing to accept the stories. This is not necessarily a positive or negative thing; the event supports their faith in relatable terms, and there is no way to disprove them.

CELEBRITY NEAR-DEATH EXPERIENCES

Celebrities aren't immune to viruses or near-death experiences. Interestingly, though many celebs have craved public attention for their real-life antics, in the NDE PR department they tend to be unusually reticent about sharing their stories, as if they don't want to be labeled as potential crackpots or drug addicts. Their stories, which the celebs generally recount after being prompted in interviews, aren't much different thematically from those told by the rest of us: blinding lights, tunnels, encounters with late relatives on the other side. They, too, received messages directing them to return to the land of the living, and many subsequently discovered life missions to complete (such as becoming a human rights activist) and/or turned their lives around by quitting drugs or finding religion. Among the roster of the famous (and still alive as of this writing) who claim to have had NDEs are Tony Bennett, Donald Sutherland, Ozzy Osbourne, Sharon Stone, Chevy Chase, and Gary Busey. Even former President Bill Clinton has said he had an NDE when his heart was stopped for seventy-three minutes during surgery. His experience was a bit darker than most: "I saw, like, dark masks crushing, like, death masks being crushed, in series, and then I'd see these great circles of light and then, like, Hillary's picture or Chelsea's face would appear on the light, and then they'd fly off into the dark."

This is precisely why it is so critical to stress the word *intelligence* in Perceptual Intelligence. It is no small feat for our unconscious minds to unscramble subconscious messages and translate them back into coherent and accurate playbacks of real life that are credible to other people. The more that NDEs are shared, the more they infiltrate our psyches, where they lay lurking as dormant puzzle pieces until the fateful day when — a long, long time from now — we face our own inevitable brushes with death.

Floating Downstream: Out-of-Body Experiences

Out-of-body experiences (OBEs) are similar to NDEs in that some reports of them also involve the patient having been dead for several minutes or more. A frequent distinction between OBEs and NDEs is that the former can happen while the individuals are still clinically alive. People have reported leaving their bodies not only during an NDE but also during surgery, while sleeping, or while in a coma. It's unusual to hear of an OBE from someone who wasn't lying down in some fashion. Most commonly, OBEs involve the individual floating upward from the body while he or she is unconscious, asleep, or dead and looking down on his or her body as if detached from it. Sometimes people see the onlookers surrounding the bed in addition to their own listless bodies.

In one famous case, a Dutch patient's dentures were removed while he was in cardiac arrest. When his nurses couldn't find the dentures after he was resuscitated, the patient told them where they were, though he was clinically dead at the time of their removal. That's a hard one to explain away.

What do we make of this phenomenon compared to NDEs? I've kept you in suspense long enough in this chapter, so I'll now recount my own out-of-body experience. This is not to convince you of any religion or supernatural malarkey but rather to demonstrate that you or others you know are not crazy if you or they have had OBEs. It also helps to provide some context for how I believe my Perceptual Intelligence was at work during these super intense and seemingly real moments.

In 1989 I was studying at Edinburgh University in Scotland. I joined the university rowing team, which involved grueling workouts in preparation for the intensity of racing up and down the rivers and lakes of Great Britain (or lochs, as they are referred to in Scotland). I was so conditioned that my resting heart rate was about thirty-one beats per minute — perhaps too low. One night in my dorm room while I was sleeping, I seemed to awake and float up to the ceiling to gaze down on my sleeping body. I had the powerful thought that I must make a choice: either I could keep moving away from my body off into the distance — which I interpreted to mean that I would die — or I could go back into my body and live. I kept thinking that I still had so much left to accomplish, and I went back down into my sleeping body. Ever since that evening I have vividly recalled those sensations as if the incident had just transpired; the experience seemed so real and was unlike any dream I have had before or since.

Before you rush to google my medical credentials, I can assure you they are genuine and have not been purchased on the Internet. I'm not crazy or deluded, and I hadn't taken medications on the evening in question. However, like the droves of others from all walks of life who attest to having had out-of-body experiences, I am convinced that I experienced something unique — and, in a sense, I did.

In early 2014 researchers at the University of Ottawa conducted a study with a willing participant who claimed to have the ability to undergo an out-of-body experience (referred to in the report as an extra-corporeal experience, or ECE) at will. Their findings, which were published in *Frontiers in Human Neuroscience*, revealed that, while the twenty-four-year-old woman was in a sleep state, her visual cortex shut off. Without the visual cortex in play, kinesthetic imagery (discussed below) was produced at a fantastic rate in her brain, during which time she had the sensation of being able to "see herself above her body rocking with her feet moving down and up as her head moved up and down as in bobbing in ocean waves." She had another ECE that "was the most intense and involved the participant watching herself above her own body, spinning along the horizontal axis."

Those Who Can't See Might Not Be So Blind

According to the *Psychology Dictionary*, kinesthetic imagery is "the cognitive recreation of the feeling of movements." If the visual cortex turns off (as with the woman in the above study) or is damaged, the result is an obvious inability to see, coupled with an unusual increase in perception. This phenomenon, sometimes referred to as "blindsight," refers to people demonstrating a measure of visual guidance without the ability to see in the traditional sense. Perhaps this is also what is behind all OBEs.

Blindsight has been known to occur, but it would be easy for us to veer too far into the territory of comics and the story of blind lawyer turned superhero, Daredevil, who loses his sight from a childhood accident, which heightens all his other senses, so let's focus on what really is occurring here. Dr. Ken Paller, a professor of psychology at Northwestern University who has written extensively about consciousness, suggests that it is possible for a person with damage to the visual cortex to "still receive visual input through projections from brain structures such as the thalamus and superior colliculus, and these networks may mediate some preserved visual abilities that take place without awareness."

Blindsight is therefore common among people who have gaps in their visual fields stemming from traumatic brain injuries. They can identify objects presented to their blind areas without being consciously aware that they are *seeing*. But what if we were to take things a step further? How about people with visual deficits navigating around obstacles they can't see and aren't expecting?

A man known to the medical world as "Patient TN" performed such a feat. TN lost the use of his primary visual cortex, the region of the brain responsible for processing the visual information that helps form conscious sight. The injury was severe enough to knock out the primary visual cortex in both the brain's left and right hemispheres, a condition called cortical blindness. Tests of TN's sight came up blank; he was unable to detect large objects moving in front of his perfectly healthy eyes. Researchers had TN walk a line without his ever-present white cane to find out how he would react. TN was reluctant, but they finally persuaded him to try. After all, how bad could it be? What TN didn't know is that the hallway was strewn with lab equipment.

Head down and hands held loosely, he maneuvered slowly but dexterously between a camera tripod and a bin, and then neatly stepped around a random series of smaller items. He flawlessly maneuvered around the obstacles, though he couldn't see a thing. About this phenomenon of blindsight, the researchers proposed not only that some aspects of vision processed separately but that vision is distinct from awareness. In other words, *seeing* and *knowing* can be completely different functions.

In the past few chapters we've explored how our minds provide insights and clues to help us unravel the world when we can't explain what our senses are registering. If our presumptions turn irrational and/or unscientific, we fall on the low end of the PI spectrum; if we question and look for answers but must relent and admit we "simply don't know," we tip toward high PI. Blind acceptance of the unknown often leads to absurdity, which high PI will call out right away.

In the next chapter we'll discuss a certain Russian ruler whose inflated ego has hijacked his PI and caused him to steal a prized ring — and attain larger conquests.

5

Vanity Games

Sand Castles, Card Houses, and the Art of Self-Delusion

In 2014 I attended the Winter Olympic Games in Sochi, Russia, to cheer on my patient and dear friend, the 2010 Olympic gold medal bobsled winner Steven Holcomb. I felt a close connection to Steven because I had helped him overcome his Keratoconus — a disease that can cause severe vision loss and that almost ended his Olympic career.

Secretly I prayed for an uneventful event, outside of watching Steven and other Americans bring home a truckload of medals. Little did I realize that while at the Games I would have a front seat to witnessing one man's quest to prove himself as the world's most powerful leader and subsequently invite everyone to join him on his massive delusion.

If there is one individual who lives in an overblown fantasy world far removed from reality, it's Vladimir Putin. Putin has little or no desire to separate reality from fantasy and in fact seems to do everything in his power to promulgate the latter, especially when it comes to his own image. His Perceptual Intelligence is so distorted he will do almost anything to convince others of his fantasy version of a "Russian Epcot Center" under his authoritarian rule — and drag millions of low PI followers along with him for the ride.

Putin's Vanity Fair

To many observers, the 2014 Winter Olympics were nothing less than a transformational event for Sochi, a palm tree–laden subtropical resort on Russia's Black Sea coast. This wasn't Russia's first top-down endeavor, but it may have been the grandest. The Sochi Olympics marked a turning point, ushering in a whole new era of the Russian megaproject. This time, however, Communist ideology wasn't the impetus but rather the singular will of the country's all-powerful leader, Vladimir Putin.

The Sochi Games were meant not only to glorify Russia but also to highlight Putin's own role in the restoration of Russian might. After seventeen years in power, surrounded by sycophants who only buttress his view of himself, Putin has a perception of reality that is more distorted than a mirror in a fun house. The personal image he likes to project suggests a vain man who believes in a superhuman version of himself: the widely reported facelift, the publicity photos showing him bare-chested and on horseback, and so on. Repeated acts of brutality at home and abroad, especially in Ukraine, suggest that Putin believes he can seize everything he desires (the shocking details await you later in this chapter about what he stole) and that nobody — not even the United States—can or will stop him. Can reality ever be brought into Putin's fantasy world?

The Sochi Olympics were Putin's Games, and he pursued them with a manic fervor, even personally attending the International Olympic Committee's final meeting in Guatemala, where the host city was being selected. While Putin didn't take his shirt off to reveal a colorful tattoo of the Olympic rings on his chest, he flexed his chest in other ways. First, he insisted that the Games take place in Sochi — Russia's southernmost point with a winter climate resembling that of northern Florida. Sochi had been best known as a subtropical seaside town and vacation destination favored by mid-level Soviet bureaucrats and budget-conscious Russians.

Why was Putin so determined to host the Games — and in Sochi, no less? Surely, as purportedly the world's second-richest man (how he achieved such wealth is a story for another time), he wasn't hard up for extra cash and could've constructed a five-star resort for himself elsewhere in his vast country, if he so desired. The purpose of these Games was twofold: they were a chance for Putin to buttress Russian superiority in front of a captive, global audience, thereby influencing their Perceptual Intelligence; and they were a statement of Putin's self-perception, which he intended to reflect outward.

These motives made sense from Putin's point of view. In the post-Communist age he was altering global perceptions of his country as something other than an recidivist, corrupt, and oppressive regime bent on propping up dictatorships, destroying the environment, slaughtering innocent animals, looting the government coffer, forcibly annexing former Soviet territories, engaging in large-scale military maneuvers, and quashing civil liberties or any obvious form of dissent.

Most important, perhaps, Putin was bent on cementing his own warped self-perception. How can you truly convince someone of something if you don't first believe it yourself? There is no doubt in my mind that Putin believes all the hype about himself. He has cultivated and nurtured his tough-guy persona, but it isn't mere bravado. He doesn't secretly see himself as a ninety-eight-pound weakling, and his image isn't just "shtick," as President Obama famously claimed. I'd argue that he believes his own press. His perception of himself and his country is an example of "wrong reality" — PI that's gone off the grid.

The Grand Delusion

In the West many fault government for almost everything that goes wrong — even events beyond government control, such as oil prices, the stock market, and Pokémon GO. In Russia, the majority of citizens do the opposite. Rather than zeroing in on

Putin for his loathsome record on everything from corruption to human rights abuses, many Russians pay blind obeisance to their strong leader, instead viewing his party and the bureaucracy as the problem and not the man himself.

How is this possible? Putin is shrewdly shaping his compatriots' PI by manipulating Russia's cultural and historic fondness for strong leaders. He has constructed his public image based on this principle, which has guided him in choosing his crusades. He has implemented conservative policies with an eye on Orthodox Christianity and statewide antiblasphemy laws, which even included the prosecution of the band Pussy Riot for protests against him and singing in Moscow's Cathedral of Christ the Savior in 2012.

Putin knew this tactic would strike a chord with people and strengthen his popularity, since much of the Russian population is conservative and religious, despite years of Communist rule. Putin's popular crusade against sodomy harkens back to the day of the czars, who were known officially as defenders of the Orthodox faith. What worked prior to 1917 was working once again. The czarist imagery of Putin's rule was clearest at the reinauguration of the Russian Popular Front in Moscow in June 2013. The Front is a political movement created by Putin and evidently intended to emancipate the president from his widely unpopular party, United Russia. When Putin entered the room, the crowd erupted in a well-rehearsed chant, "People, Russia, Putin" — a slogan clearly derived from the czarist "Autocracy, Orthodoxy, Nationality," which described the system of political legitimacy before the 1917 Communist Revolution.

SELF-IMAGE ON STEROIDS

There are some striking similarities between Putin's reign and President Donald Trump's administration, as well as Trump's influence on how others perceive him. This is not so surprising, since Trump has a long history of documented public and social media comments reflecting his admiration for Putin. In 2007 he told Larry King he thought Putin was "doing a great job in rebuilding the image of Russia and also rebuilding Russia period." In 2013 he tweeted twice about Putin, once wondering if the Russian president was going to be his "new best friend" and another in which he pondered whether he would attend the Miss Universe Pageant in Moscow. During his presidential campaign, Trump seemed overjoyed that Putin had referred to him as a "genius." Putin and Trump met face-to-face for the first time in the summer of 2017 in front of the press corps. They appeared rather chummy. It was like a scene straight from *Goodfellas* or *The Sopranos* when Putin leaned over to Trump, pointed to the journalists, and asked Trump, "These are the ones hurting you?" Trump replied, "These are the ones."

Was President Trump consciously (or perhaps subconsciously) following Putin's lead in order to create a specific self-image and influence our Perceptual Intelligence? Possibly. The US president, like Putin, is deeply concerned about his self-image and has used similar techniques to create a perception of "strong leadership" to the public (although we haven't as yet seen our president riding bare-chested on a horse). President Trump was essentially saying that he is a "great and strong leader" by virtue of continuous association with Putin's praise and admiration for him.

The Sochi Aftermath

When Russia first bid for the 2014 Games, the cost was projected at a relatively trivial $10 billion (well, as far as the Olympics go) — and landed in the ballpark of a staggering $51 billion. To curb government spending, Putin passed the Games' costs on to the country's top billionaires. But spending spiraled out of control for even the richest Russians, and the state engaged in a high-level Ponzi scheme, using one of its own banks to loan the billionaires cash, which they'd then donate back to the government. The Russian government initiated twenty-seven separate "official" criminal investigations into alleged embezzlement, but not a single soul was ever brought to trial.

And that's not all. Sochi's gleaming stadiums and state-of-the-art venues were completed only after thousands of migrant laborers worked around the clock and under horrible conditions, with virtually no days off. Anyone who complained was fired or deported.

Sochi's three hundred thousand–plus residents didn't fare much better. They endured daily power outages, months without running water, and construction-site pollution. An estimated two thousand families were evicted from their homes by the Russian equivalent of eminent domain and, despite a promise of the greenest Olympics ever, thousands of ancient forest acres were slashed in protected Sochi National Park, as well as at a nearby UNESCO World Heritage site.

It's now widely known that Russian officials carried out a doping program at the Sochi Games by giving their athletes performance-enhancing drugs and then tampering with their urine samples to cover it up. The doping involved at least thirteen Russian medal winners, including Alexey Voevoda and Alexandr Zubkov — Steven Holcomb's primary competitors in the bobsled competition. Russia won more medals than any other country, and not a single athlete was busted during the games.

In 2014 the World Anti-Doping Agency investigated the Games and issued a report identifying Grigory Rodchenkov, the man who handled the testing for thousands of Olympians, as the linchpin. Russia was provisionally suspended from international track and field competition, and many athletes were banned from competing in the 2016 Summer Games in Rio de Janeiro.

For all its fanfare, the Sochi Olympics are a chilling example of how one man sought to solidify his public image at all costs — and, to the extent of his influence, succeeded. Two years later, in May 2016, Putin achieved an astonishing 82 percent approval rating, which included positive support from opposing parties. With all the blatant corruption, misspending, and malfeasance at Sochi, how did Putin, who was so embedded in the landscape of the Games, not only get off scot-free while others took the fall but advance his public image of strength to gargantuan proportions?

THE ART OF THE TURNOVER:
PUTIN STEALS KRAFT'S SUPER BOWL RING

Never before has the term *political football* been more apt than when referring to the bizarre relationship between New England Patriots owner and billionaire Robert Kraft and Russian president Vladimir Putin. The story begins in 2005, when Kraft visited Konstantinovsky Palace in St. Petersburg with Citigroup chairman Sandy Weill and met with Vladimir Putin. News reports afterward stated that Kraft had shown Putin his Super Bowl XXXIX victory ring, whereupon the Russian leader admired it, pocketed it, and walked off. Further stories hinted that Kraft hadn't been a willing gift giver. Several months later, Kraft released a statement in which he said he "decided to give him the ring as a symbol of the respect and admiration that I have for the Russian people and the leadership of President Putin." In June 2013, while speaking at an event in New York, Kraft provided a revised account of the incident and accused Putin of being a ring thief: "I took out the ring and showed it to [Putin], and he put it on and he goes, 'I can kill someone with this ring,'...I put my hand out and he put it in his pocket, and three KGB guys got around him and walked out." Kraft was so upset by this that he took up the matter with the Bush White House, who replied it would be in the best interests of political relations to treat the ring as a gift.

After this version of the story hit the press, Putin said he didn't remember meeting Kraft and offered a "proposal": "I'll ask our firms to put together a really good, big thing, so everyone will see what an expensive thing it is, with good metal and a stone, so it will be passed from generation to generation in the team, whose interests are represented by Mr. Kraft." When Putin was asked again about the incident, his official spokesman implied that Kraft was delusional and might need "psychoanalysis." Though the situation remains in dispute, Kraft's 2005 Super Bowl championship ring continues to be in Putin's possession. Today it's showcased in Moscow in a Kremlin library that is reserved for gifts. It's not known if Kraft has ever gone back to visit his ring.

The Reality of Influence

Poll after poll shows that the Russian people prefer a strong ruler to a solid democracy. While this may appear baffling, it makes perfect sense; centuries of Russian history and cultural conditioning have molded their perceptions of what a leader should be and how this individual interfaces with the government and the people. Putin is well aware of this and has shrewdly influenced his country's Perceptual Intelligence; he has created and sustained an image of himself that taps into the people's collective reality of what a leader should be.

Over the years the list of Putin's well-photographed macho endeavors has included fishing, deep-water exploration, horseback riding, motorbiking, firefighting in the air, scuba diving, rafting — and even sedating a tiger. To accentuate his image at charity events, he has sung, played piano, and contributed his original paintings. Talk about a Renaissance man!

By contrast, it's hard to imagine a US president going to such lengths to convey a public image of this magnitude. Theodore Roosevelt may have relished hunting, but this was a singular personal passion; while the image did appeal to a certain frontier aspect of the American people, this is simply who he was and not a facade he conjured up and sought to promulgate. In June 1992 Bill Clinton donned sunglasses and played saxophone on *The Arsenio Hall Show* while running for president; although it helped win some voters over by presenting a fun aspect of the candidate, it was more of a publicity stunt than a long-term image creation. As president, he didn't repeat his sax solo on any other late-night television show.

Putin's fixation on his self-image is so all consuming that he has a need to reinforce it at every turn. Whether it's through the might of the Olympics, being photographed as an outdoorsy macho man, or getting the upper hand with an American billionaire by stealing a symbol of American athleticism, Putin has created a fantasy vision of himself that is now widely accepted by the majority of his people. He has tapped into the collective unconscious of Russia and given them the image of a leader that has been planted deep in their psyches, going back to the czars.

Does Putin see himself as he really is? Or has he distorted reality to such an extent that he now believes the facade he created? With no parents around to figuratively smack some sense into Putin, I think he believes his manufactured image is real, but we may never know for sure. We can be fairly certain, however, that for as long as Putin can maintain his outward image, he will have the support of his government and his people, no matter what else he does because they *like what he represents*. On a subconscious level they feel they will be a strong nation as long as they have a visibly strong leader at the helm.

The Russian conqueror lives in a fabricated world with delusional PI. Putin may not be a professional-caliber athlete by any stretch of the imagination — though he obviously would like to be seen that way, as evidenced by 2017 photos of him in hockey gear — but there are genuine standouts in the sports world who have tapped into their high PIs to achieve greatness, as well as stars with such low PIs that they plummeted from grace. We'll cover this and the perceptual hazards of fan fanaticism in the next chapter.

6 Let's Get Physical

PI and Sports

We marvel at athletes who can complete the New York City Marathon in two hours and twenty minutes, smash a baseball five hundred feet over a fence, or ski down a slope at eighty-five miles per hour. Whether he or she is performing in an individual competition, in a team sport, or at the Olympics, it takes far more than good genetics, physical prowess, hand-eye coordination, and hours of intense labor and practice for a professional athlete to become number one in his or her field. Those things certainly come in handy, as do more difficult-to-measure factors, such as perseverance and determination, but when success reaches the level of Michael Phelps or Serena Williams, something else is at play — and it's not anabolic steroids.

Once champions reach the top, they seem unflappable, unstoppable, and unsurpassable when it comes to confidence and winning — but how, exactly, did they get there? Did they eat their Wheaties every day? Gulp down five raw eggs à la *Rocky*? Or did they have "lucky" tricks, like Michael Jordan sliding into his blue North Carolina trunks before donning his Bulls uniform?

Whether or not these athletes realize it, their Perceptual Intelligence has been there every step of the way — albeit in different forms, depending on the circumstances and the individuals. In this chapter I'll explain why it's so crucial for anyone who participates in sports and exercise to have high PI in order to make the leap of faith and be convinced that all the hard work is worth the pain and effort. We'll also unravel why PI causes some superstar athletes and teams to fail and others to skyrocket to the pinnacle of their sports, only to plummet straight to the bottom again. And last, we must not forget the dedicated sports fans, whose emotions, belief systems, actions, and tattoos are significantly affected by their teams' day-to-day and year-to-year performances and cause them to lead fantasy lives due to their low PIs.

Giving Your Brain a Noncaffeinated Jolt

We've been told for decades that exercise is good for our bodies in terms of weight management, cardiovascular health, muscle strength, flexibility, stamina, sex, and life span. Equally important is how physical activity interacts with and stimulates the brain in innumerable seen and unseen ways. Numerous studies have shown that exercise can help reduce stress, improve mood, and even counter the impact of negative thinking and depression. The residual effects, scientists believe, also include heightened self-esteem and a decrease in the perception of pain. So a day or two after your next gym workout, you can focus on all these positive benefits as you are popping Advil and rubbing your aching arms and legs!

How the brain garners these benefits is no longer a mystery. During exercise, the body unleashes chemicals known as endorphins that travel to receptors in the brain, stimulate activity, and cause what some people perceive as a feeling of euphoria or as a "high." In fact, some of the endorphins are also known to have an analgesic impact on your brain and spinal column, much like morphine.

Having said that, I recognize that the power of science probably isn't enough to inspire you to start exercising if you've grown roots into your couch cushions. The reality is that many of us abhor exercise, despite acknowledging its benefits. Something contrary occurs when we haven't exercised for a while: our brains turn complacent and lazy, we feel lethargic, our muscles atrophy, and we dread putting on those workout shorts. Ahead of time our minds visualize a workout overwhelmingly filled with images of agonizing and unnecessary exertion: the time-consuming drive to the gym, the dreary warm-up, the annoying wait to get on the right machine, the crackling of formerly hibernating joints, the lactic acid burn of muscles, the desperate gasp for a next breath, and the sweat on the forehead. The muscles haven't been used for so long, they tell us "nah," and we make up all kinds of excuses why we couldn't or shouldn't go: "I'll flare up my old football injury," "I won't get back in time to watch *Dr. Phil*," "It's too cold outside," and on and on until we plunk down on the couch with the remote in one hand and a jelly donut in the other. Our endorphins lay inactive, and we feel sleepier than ever before and intensely guilty about not having gone to the gym. "Tomorrow," we think. "I'll do *double* tomorrow."

If you are one of those chronic procrastinators whose January gym membership never saw usage after February 1, somewhere deep down you know that heading to the gym is not on your calendar for "tomorrow" — or the next day or the day after that. Why? Because your mind has now associated exercise with all your negative thoughts, reducing your PI by causing you to accept a distorted view as fact.

On the flip side, if you were to work out once a week, then build up to twice, and go all the way up to five or six times, something else would happen in your mind. Instead of picturing yourself on a treadmill clutching your chest, you start thinking about a bunch of other things: how much weight you've lost, how much better your body looks to you in the full-length mirror, and how you might be interested in a new dress or suit to fit your svelte new body. Some of us become so consumed with exercise that heading to the gym becomes a compulsion. Why? Because it felt good during and after that state, and we are at last relating exercise to memories and images of positive sensations and rewards.

Now imagine that you are a professional or elite athlete who undergoes several hours of grueling exercise and training every day. The motivation to push harder, faster, and longer is not just about the aforementioned benefits; it's about your livelihood, reputation, family, legacy, sponsor, intense lifelong desire, and the trials and tribulations of past victories and defeats. The brains and bodies of these athletes are conditioned from repetition — and the urging of coaches and trainers — to know what a rigid routine of stringent exercise truly means, and they have formed pictures in their minds of what the end results look like. In this regard, PI becomes a road map guiding athletes through their journeys and reminding them both consciously and subconsciously of why they must continue to work even harder every day.

For professionals, elite athletes, and exercise addicts, incessant physical activity has one additional effect on the brain: it improves white matter activity. You may have heard of gray matter — as in "the old gray cells ain't what they used to be" — which is where we process information and where the senses are controlled. White matter, which you hardly ever hear about, is approximately *60 percent of the brain* and serves as the connective tissue or communication system for the gray matter. Current scientific belief is that if we don't exercise enough, the white matter begins to decay, and over time we suffer memory loss, and our ability to process visual and auditory information becomes hindered. Once these functions are affected, our willpower gets wildly diminished, hindering our PI; we can't remember details, so we focus on old perceptions and stubbornly stick to them. How often have we seen an elderly couple bicker over a mundane faint memory, such as whether Aunt Fern's long-dead cat was named Sam or Max?

With frequent exercise over time, however, white matter has greater staying power, which means improved memory, sharper perceptions, and less risk of dementia and Alzheimer's. A recent study revealed that white matter among physically active people in their seventies was superior to that of sedentary groups of individuals — often if the latter were younger, too. In terms of PI, the sky's the limit for people who exercise consistently over time;

since their white matter and connective tissue are well preserved, they will tend to be more accurate and focused when it comes to interpreting reality and telling it apart from fiction. Whether it's something that occurred years ago or a live incident witnessed in real time, these individuals will likely be able to visualize and describe the event with razor-sharp precision. Chances are they will be able to relate to and convey the emotions experienced during those moments with bull's-eye accuracy as well. That alone is worth fighting off thoughts of muscle cramps and chest heaves to make a beeline to the gym, isn't it?

Sports Is All about Confidence and Perception

Having high PI in sports is all about how you see yourself fitting into the landscape. Do you feel relaxed and confident about your playing abilities on the field? Or does your inner voice tell you, "Oh no, last time that pitcher struck me out four times — he's going to make me look bad again, I know it." In that case, you've lost before you've even grabbed the bat.

In one experiment, twenty-three volunteers were asked to kick a football through the goalposts from the ten-yard line. After a warm-up, they were asked to judge the height and width of the goal by adjusting a handheld, scaled-down model of the goal made out of PVC pipes. They then each performed ten kicks. Immediately after the final kick, participants repeated the perceptual measurement.

The result was striking. Before kicking, both groups had the same perception of the size of the goal (incidentally, an inaccurate one). After ten kicks, the poor performers — those who scored two or fewer successful kicks — saw the goal as about 10 percent narrower than they had before and perceived the crossbar as being too high. On the other hand, the good kickers — those who scored three or more — perceived the goal to be about 10 percent wider. In other words, how well you perform influences the way you see the world and determines your PI.

In another study, elite female rowers were tested on a rowing machine known as an ergometer. One group was told by the researchers that they were going faster than they actually were. Another group was told that their pace was slower than it was. The results tell us a fascinating story: the former group slowed down their speed because they thought they couldn't maintain it, while the opposite occurred in the latter group — they *increased* their speed to catch up. On both teams, perceptions were off due to incorrect information, which had a direct impact on their performance.

Our minds have been known to help us out when we're feeling lousy about our performance or in need of some justification for it. A kicker who misses an easy extra point might make the excuse that the wind was high — and become convinced of it as fact. But there is an evolutionary purpose to this adaptation. If someone perceives a goal as higher or smaller than it actually is, he or she corrects for this mistake and aims more accurately the next time around.

Assuming we're of sound mind and have not sustained neurological damage, our conscious perceptions of the world are stable. Strength levels, confidence, fears, desires, and any number of mutable factors fuel our awareness of the objects around us. Yet there are a number of things athletes can do on a subconscious level to influence their minds to help their bodies perform better.

Thinking Visual

In today's sports world there probably isn't a single professional athlete who hasn't tried at least some version of *self-visualization* in order to perform better. On the surface it may sound like some kind of hocus-pocus, but self-visualization has been an effective device for many athletes in a variety of sports. In layperson's terms, self-visualization is a technique used to create mental images that help turn innermost thoughts, desires, and goals into reality. Self-visualization is used not only by athletes; it can be applied by anyone about to embark on an endeavor — from an employee vying for a corporate promotion to an actor auditioning for a plum role.

Self-visualization isn't as much wishful thinking or dreaming ("If you build it, they will come") as it is a way to improve focus, memory, and perception. In baseball, for example, hitting is a repetitive skill that requires a proper batting stance, a smooth and level swing, lightning-fast wrists, and superb hand-eye coordination, among other things. If you picture yourself swinging the bat correctly and with authority over and over, your brain will translate that visual picture and store it for when it's time for you to saunter up to the batter's box. Many self-visualization exercises involve envisioning the end result, such as the ball landing over the fence, the ball swooshing into the basket, the puck landing in the net, and so forth. By imagining the successful execution of the activity, the athlete is also producing a fantasy to process those movie frames and play them back anytime they're needed; the fantasy becomes the reality because the mind has seen it as happening and is signaling to the body how it's done, which boosts confidence. How can it not be destined to occur when you've played it over and over, with the desired success lodged in your head?

I had the privilege of consulting with my friend Hall of Fame tennis pro Pam Shriver on the subject of self-visualization. While she conceded that self-visualization wasn't standard practice back in her heyday in the 1980s, she did a form of visualization "*while* playing tennis more frequently than [she] did before the match." Shriver was well aware that her strength and her bread-and-butter lay in her serve. "My serve was the most important shot in my repertoire. If my serve was off, I suffered more than any other," she described. "Some [tennis players] might visualize technique, but I would visualize the serve and the placement of the shot and where the ball would bounce....I became so confident in my serve that I could hit a quarter."

Getting in the Zone

We've established that strenuous exercise over the long term fires up the white cells in the brain and that self-visualization helps put winning images into our mental movies, but another critical factor has been found to transform an athlete from good to great to legendary status: getting into "the zone," the athlete's brass ring. Male and female jocks talk about it as if it were a real-life place to visit, a sports utopia where only the truest of professionals and elites are entitled to reside. When athletes refer to "being in the zone," they describe how everything happens with ease and naturalness; it's as if the heroics that occur were *meant to be*. In the sports zone, the hitter's bat feels lighter and the ball seems larger, appearing to be coming at him or her in slow motion — just like in the climactic scene in the film *The Natural* (based on Bernard Malamud's novel). This isn't reserved just for major league baseball players: in hockey, the net seems wider; in basketball, the hoop appears lower; in skiing, the slopes feel like clouds. We're most mesmerized by athletes when they reach the zone because it feels as if we're watching poetry in motion and, in those moments, it seems nothing can stop them. For these athletes, this zone has an aura surrounding it, as if it had been preordained by destiny. Is the zone just a cliché, or is there a clear path to getting there? Can one just download Google Maps into the brain to enter this golden place?

There is perhaps no one in the history of professional sports who has resided in this golden place more than retired NBA star Michael Jordan. He owes at least some of the credit to sports psychologist and meditation authority George Mumford, who also authored the book *The Mindful Athlete*. While there are elements of self-visualization involved in Mumford's training with NBA stars — such as imagining the basket before taking a shot and anticipating what happens even before it lands — Mumford takes the concept leagues further by guiding athletes on how to "be in the moment," using various meditation techniques he learned from Jon Kabat-Zinn. This is no easy skill to hone; it's not in any way like learning to dribble or shoot. It takes endless hours of time, patience, and repeated meditation sessions, and the athletes require not only intensely sharp and focused minds but the ability to turn off all other thoughts at will. That sounds counterintuitive and off-the-wall absurd. Why on earth should NBA or WNBA players flip off their conscious minds in the middle of a game? Wouldn't the other team run circles around them?

Yet by being able to turn off the mind, athletes can head straight toward the subconscious. This means bypassing all the head noise — a squabble at home with a significant other, the steak for dinner that night at a fancy restaurant, the journalist's snarky comment in the newspaper the day before, the deafening jeers from the crowd — in order to connect images formed over and over again during meditations.

For Michael Jordan, "being in the game" meant being able to dip into those deep intuitive reservoirs at will and knowing precisely which move to make in a tiny fraction of a second. For a man with off-the-charts high PI (in terms of creating a fantasy image and then fulfilling it as reality), unparalleled intuition, and enormous competitive drive like Jordan, no conscious deliberation was necessary. He could simply get his mind to snag the right saved images and apply them at the exact right time. Since his conscious mind had been cast aside, his subconscious took over and formed an easy conduit between his nervous system and his muscles, with nothing hindering this process. In this way Jordan could get in the flow in the spur of the moment, gliding through the air and over and around opponents as if possessed.

Many contemporary athletes have found their own methods for creating the reality they seek using PI. Billiards player Will DeYonker, who is on the autism spectrum, has a gift for seeing 3-D images in his head when attempting a complex trick shot, which he maps out in diagrams that he carries to tournaments and events. Renowned big wave surfer Laird Hamilton, who has surfed seventy-foot waves, speaks extensively about how he gets in the zone. Multiple Olympic gold medal winner Kerri Walsh Jennings, who has won more beach volleyball tournaments than anyone, has been spotted both meditating and doing brain training. Steven Holcomb, the most celebrated bobsled driver in US history, could see himself driving every turn on every bobsled track in the world. Tommy Pham of the St. Louis Cardinals baseball team, a patient of mine with Keratoconus, focuses, while he's on deck to bat, on seeing the ball come out of the pitcher's hand and visually tracks it toward the batter. It all goes to show that though being in prime shape goes a long way, tapping into the deep recesses of the mind by visualizing success has the potential to hurtle a great star into the stratosphere.

Players also use these techniques to break up the flow of their opponents. Pam Shriver admits that she, like all tennis pros, faces days when she is "off" and the woman on the other side of the court is eating her for lunch. "It could be that you have a virus or you just don't feel right," she describes. "There are little things you can do. You can call time and sit in your chair to break up the other player's flow." In other words, if you are facing a seemingly invincible opponent who is in the zone, you have to find a way to shut down the film in his or her mind. By breaking your opponent's concentration, you're separating mind and body and bringing that person back to full consciousness, where distraction and doubt are brought back into play. When it comes to defeating and lowering the PI of professionals at the top of their game and in the zone, it takes real champions to find the courage, concentration, and strength to battle back with their own resourcefulness and zone training.

Being in the sacred zone is not reserved just for professional and elite athletes — or even sports. We all have been in the zone to a degree in some area. When I do eye surgery I am so focused on my work that I don't even notice if there are TV news cameras or observers hovering around me. It's as if these objects and people don't exist. My high PI blocks everything out and my perceptions are solely of what's in my microscope ocular lenses. Have you ever been so engrossed in reading a book that you didn't hear someone call your name? The zone is available to most everyone, and practicing focus to enter at will and remain there can raise your PI in the process.

Teamwork and Winning Are All about Reading PI

We often hear about couples who can read each other's minds and anticipate what the other is about to say or do before he or she does it; that happens from time to time with my wife and me. Shriver has a lot to offer on this subject. Although she was an excellent tennis player, her abilities soared when it came to playing doubles. She won a whopping 112 career doubles titles, seventy-nine of which were played with none other than the great Martina Navratilova. Now, one could attribute Shriver's longtime success solely to Martina — the game's most dominant player at the time — but how does that account for the thirty-three wins Shriver had with other doubles partners? And why would the game's most accomplished player choose a partner if she didn't think she was qualified (which she obviously did)?

While building a winning team, you obviously need to first select the right players and a solid coach. After that, every team member requires the discipline to undergo taxing strength training, drills, and practices in preparation for game day. Coaches will lecture about "teamwork" and "good communication" among players, but there are some things they can't control on the field: synergy, rhythm, and constant connection. Shriver recalled her unique on-court link with Navratilova: "It was very intuitive what we were doing. We seldom had mixed signals. If there was a lob over my head, I knew she would be there to get it. We would know what the other person was going to do. We could read each other's moods."

Shriver's level of confidence about reading and reacting to

her teammate is astonishing. It's more than just trust in ability or talents meshing. I believe that Shriver and Navratilova, like members of all great duos and teams, had the ability to seemingly get into each other's heads. The words Shriver used — *intuitive* and *read each other's moods* — are about picturing how her teammate will react to a variety of circumstances. Shriver had stored in her head film clips from eight and a half years' worth of doubles matches and was able to play them back in that same fraction of a second that Michael Jordan did while floating in the air toward the basket.

The greatest sports teams of all time — from the 1927 New York Yankees to the 1976–77 Montreal Canadiens to the 1998 Chicago Bulls — have had that rare mix of physical ability, leadership, durability, and teamwork. They each have had team members who could complement one another's skills and temperaments (think of the yin and yang of Yankees' greats Babe Ruth and Lou Gehrig). In terms of PI, all these teams have had rare hyperconfidence — like that of Navratilova and Shriver — in which the entire team shared the vision of winning every game. In fact, the 1927 Yankees' winning was such an expected, foregone conclusion — the team won 110 games out of only 154 played, 19 more than the second-best Philadelphia Athletics — that legend has it their World Series opponents, the Pittsburgh Pirates, were intimidated beyond recognition simply by watching the Yankees take batting practice the day before the first game. (The Yankees went on to sweep the series.)

This shared winning contagion isn't just a matter of talent or cockiness, though those things are certainly factors. The University of Connecticut women's basketball team (UConn) — which has won an astonishing eleven NCAA division championships (so far!)* — has no shortage of both these things, but it has repeatedly claimed victory with a slew of talented players over a lengthy

---

* The UConn team won its hundredth *consecutive game* as of February 14, 2017.

stretch of time. Certainly the team has a knack for continually recruiting the best talent, but Coach Geno Auriemma knows how to keep all egos in check and play to everyone's strengths. His expectation is the mind-set of constant winning: "Fair or unfair, it's not good enough just to win. There's a perception that if we don't get to the Final Four, it's a bad season." The team has become so powerful it's even led to some journalists accusing the Huskies of being "bad for the game." They are such a force that the Springfield, Massachusetts, newspaper the *Republican* refused to print the game box scores because, in the words of journalist Ron Chimelis: "To me, that's like printing the final vote of an uncontested election."

LOVABLE LOSERS: SHARED PI AT ITS WORST

Just as winning outlooks are often ever-present on a successful team, the reverse can also be true when it comes to an unfortunate assemblage of players: the picture of losing gets so embedded in the team's mind-set that it becomes contagious and the players don't do anything right. Many of us can picture the comical gaffes of the lovable losers in *The Bad News Bears* films, for example. Year after year, the Philadelphia 76ers NBA basketball team is a real-life example of this.

The 1962 expansion New York Mets were the real deal when it came to contagious awfulness. They were so dreadful that their 120 losses don't begin to describe how pathetic they were. Among their claims to shame were 210 errors, a .240 team batting average, two twenty-game-losing pitchers (a third, Jay Hook, lost nineteen!), and a Dodgers pitcher, Sandy Koufax, throwing his first no-hitter against them. Fittingly, they lost the last game of the season by hitting into a triple play that squashed a potential eighth-inning rally. It took seven long years, a pitching franchise named Tom Seaver, and the managerial skills of Gil Hodges to reverse the mind-set and create a professional environment, leading to the 1969 "Miracle Mets" team.

Fallen Stars

Sometimes it seems as if the greater the accomplishments and the higher the profile of the male athlete, the more sordid the headlines. The roster, in no particular order, of all-star talent in a range of sports who fell from grace owing to a variety of accusations is mind-boggling: murder (O. J. Simpson), gambling (Pete Rose), steroid usage (Mark McGwire), doping (Lance Armstrong), substance abuse (Lamar Odom), rape (Mike Tyson), extreme infidelity and promiscuity (Tiger Woods), sexual assault (Kobe Bryant and Lawrence Taylor), dogfighting (Michael Vick), and so forth. Why do so many illustrious athletes have a hard time keeping their noses clean?

Male athletes are molded to have a certain outlook. At the risk of overgeneralizing, I believe that they are constantly pumped up by club and high school coaches, parents, and fans, all of whom encourage their aggressiveness. For years their minds show them a film of winning and a self-image of perfection and superiority that encroaches into other areas of their lives. The high PI in terms of winning causes low PI in terms of real-life; in other words, some athletes live in the fantasy that they are immune to social mores and that they can get away with anything.

There perhaps was no other athlete in sports who motivated his team like Hall of Fame football player Joe Namath. The handsome quarterback cockily predicted the New York Jets' 1969–1970 Super Bowl victory against the mightier Colts and, as promised, handed them one of the greatest upsets in sports history. Yet

Namath, an idol to a generation and something of a sex symbol in his prime, had a PI that associated his gifts as a quarterback with the spoils of success — mostly women and alcohol. Dubbed "Broadway Joe" for his stylish flair, a drunken Namath was caught on camera making an inappropriate advance to reporter Suzy Kolber that no sports fan is willing to forget.

The fall from grace for Namath and for all these other seemingly immune superstars was steep. The availability of money, alcohol, sex, and drugs is too tempting for some male athletes, whose emotions are raw and interwoven with their low PIs and their high testosterone. These athletes, who have convinced themselves of their athletic invincibility, can't separate the instincts that made them brilliant on the field from their behaviors in their private lives. Their self-perception of invincibility devastates their PI outside their athletic talents, and allows them to make poor choices that come with severe consequences. PI can giveth and it can taketh away.

Over the years, a few athletes have been able to redeem themselves — to at least some fans. Mark McGwire, who admitted to having taken performance-enhancing drugs, was allowed back into major league baseball as a bench coach for the St. Louis Cardinals. Mike Tyson has received praise for his film work (in *The Hangover*) and has, at least on the surface, become something of a reformed family man. After serving prison time for animal abuse, Michael Vick has paid off his financial debts and returned to the NFL. He now does PSAs to raise money for abused doe-eyed puppies. (Nope, he doesn't — I made that one up. I'm a huge dog lover.)

For others — like Pete Rose and his never-ending quest to be inducted into the Baseball Hall of Fame — forgiveness and redemption have been elusive through no one's fault but his own. When the perceived integrity of the game is at stake and teammates are affected, the consequences tend to be more severe and lasting. It's doubtful that Lance Armstrong, who won the Tour de France an unprecedented seven times, will ever be able to reclaim anything close to his former standing. Not only does the cycling world feel that Armstrong betrayed, undermined, and poisoned the sport by doping, but there is the quadruple-whammy of his having pressured teammates to partake as well. In this instance, the shared PI state of winning became so frenzied and distorted that others felt compelled to follow Armstrong's·lead — and many of those involved paid a stiff price for getting sucked into the same vision.

Fan Fever: Interwoven in Our DNA and Jerseys

We've seen fans at stadiums and even at home paint their chests, dye their hair, and tattoo their bodies to convey to the world their wholehearted love and devotion to their teams and players. Fans, like the athletes they cheer and jeer, have long-ingrained memories from their childhoods of being in the stands wearing team jerseys, along with their older siblings, parents, and grandparents. Comedian Jerry Seinfeld has joked that we are "cheering for clothes," not professional teams or players, because both change cities so regularly, yet we still root for new players wearing the same team uniforms. Generations of New York Yankees and Red Sox fans — as well as the manic football enthusiasts in Green Bay, Pittsburgh, and Oakland — would be inclined to disagree, going so far as to say that the team spirit and colors make up the fabric of their DNA. Such team dedication is by no means relegated to American sports: the UK football (soccer) fanatics in Manchester and Liverpool, England, break out into all-out pub brawls on hearing the slightest peep of disparagement against their respective teams, being seemingly more sensitive to that than to insults to Mum or Dad. And you wouldn't want to get in the way of hardcore Montreal Canadiens fans when they celebrate an important win with a riot. In these cases, fans' perceptions are so tied to their emotions that their distorted PI can lead to grave consequences. Remember: *fan* is short for the word *fanatic*.

If you're a New York Yankees fan, you are drawn to a legacy

of pinstripes, championship rings, fall heroics, and larger-than-life Hall of Fame–caliber players: think Babe Ruth, Lou Gehrig, Mickey Mantle, Yogi Berra, and Reggie Jackson. Most recently added to the canon are Derek Jeter and Mariano Rivera. The Yankees organization is well aware of this tradition, which is ingrained in the imaginations of their fans, and naturally plays to it in advertising, merchandising, and the cost of stadium seats in the Bronx.

Teams who have experienced title droughts that seemed to last forever — such as the Boston Red Sox and the Chicago Cubs — maintain a plethora of fans with unbridled team pride even during the worst of times. But for years these fans' minds were laced with images of curses, balls going through players' legs or interfered with by fans, and crushing late-inning fall losses. Before the Red Sox and Cubs had their long-overdue moments with destiny, the fans suffered in unthinkable, tragic ways; these moments affected their PIs, causing intense heartache that leaked into their relationships and careers over many decades.

In many cases, being associated with a losing team meant low self-esteem and even depression. For eighty-six years, Boston Red Sox fans suffered terribly, until the Sox finally broke the "Curse of the Bambino" (Babe Ruth) in miraculous fashion in 2004 by defeating their evil foes, the Yankees, in the American League Championship Series and then sweeping the Cardinals in the World Series. It could be said that Boston Red Sox fans have since been seeking all-new identities not associated with failure; now that they can boast of having a team with three recent World Series trophies, they no longer identify with the woe-is-me perspective. Sales of antidepressants might now be at an all-time low in Boston (especially given the continuous success of the New England Patriots!).

There have certainly been cases of extremist behavior among fans whose mind-sets are way too wrapped up in their teams and in winning. The fans who overturn cars after a win (or a devastating loss) are fanatics in the most dangerous sense of the word.

All the extreme fans' emotionally charged hopes and dreams have rested on winning and have built up inside until they burst. Death threats awaited Steve Bartman, the Cubs fan who interfered with a ball during a 2003 postseason game at Wrigley Field.

Perhaps the most tragic story of out-of-control fan-based PI took place at a Colombian bar in 1994, where soccer player Andrés Escobar Saldarriaga was shot no fewer than *six times*. The player was killed after having cost his team the FIFA World Cup by having accidentally kicked the soccer ball into his own goal. The three men associated with the killing could have had cartel affiliations and a vendetta related to gambling losses, but there is no doubt that fanatical team and national pride were behind the *Godfather*-like murder.

If you think billions of fans around the world see their teams through wildly distorted PI lenses, just turn the page, and you will discover how and why scores of people believe they see some pretty freaky things on their dinner plates.

7 Immaculate Perception

Would You Pay $28K for a Grilled Cheese Sandwich?

How much would you be willing to shell out for a grilled cheese sandwich? $6? $7? Maybe $8, tops.

In 2015, at New York City's popular Serendipity 3 restaurant, sober diners line up for the dubious honor of plunking down $214 (not including tip) for the Guinness World Record–winning Quintessential Grilled Cheese Sandwich, which features bread made from Dom Pérignon champagne and edible twenty-four-karat gold flakes, slathered with white truffle grass-fed butter. (Serendipity currently charges $1,000 for its Golden Opulence Sundae, which requires forty-eight hours' notice to prepare…but that is a story for another book.) So much for starving kids in China.

Serendipity's Quintessential Grilled Cheese Sandwich paled, at least in price, to the partially eaten grilled cheese sandwich formerly belonging to Floridian Diana Duyser, who, in 2004, auctioned off her remarkably well-preserved ten-year-old sandwich for a whopping $28,000. What could possibly have made that sandwich worth so much to anyone? If you'd like to see the sandwich, here's where you can find a photo: http://content.time.com/time/specials/packages/article/0,28804,1918340_1918344_1918341,00.html.*

---

* If you successfully typed this ridiculously long domain name without a single error, then I commend you and you should consider a career as a court room stenographer (reporter).

This was no ordinary grilled sandwich, and Duyser is not your run-of-the-mill griller. The Miami resident claimed that she saw an image of the Virgin Mary emblazoned in the pattern of the sandwich's skillet burns. Odd as it may seem, there was enough buzz surrounding this newly anointed holy grail of grilled cheese to generate a trending global news story, piquing the interest of a deep-pocketed, if slightly unhinged, horde of eBay bidders. Duyser's godly creation, which she claimed never sprouted a single spore of mold over all those years, joined the ranks bestowed on such other consecrated food items as the Mother Teresa cinnamon bun ("nun on a bun," as it's now affectionately known), Jesus Cheeto (or "Cheesus," if you prefer), and the Prince of Peace pierogi.

Over the years people have tried to sell some pretty ludicrous items. In 2003 Sir Paul McCartney's cold germs were offered up at auction. A fan intentionally spent time next to Sir Paul when he had a cold in order to contract it himself, which he did; he coughed into a bag and sold the musically inspired mucous on eBay. A year later, soccer player George Best's liver went up for sale after a transplant (though some reports indicate it may not have been genuine — but who could really attest to the difference?). In 2005 a home pregnancy test used by Britney Spears in a hotel room sold for $5,000. Just imagine the dinner-party conversation.

These "sacred" mementos did not involve faces in food, as did Duyser's grilled cheese, but rather were hyped as being connected in some tangible — and, frankly, gross — physical capacity to contemporary celebrities. But what was it about this particular Virgin Mary grilled cheese sandwich that caused such a stir and commanded the inflated price? Are all these people — sellers, bidders, collectors, and buyers — kooks?

As we've established, our Perceptual Intelligence bridges gaps of interpretation, helping us discern whether the image reels playing in our heads are legitimate. Our deep-seated experiences, backgrounds, and cultural influences weigh in, helping us draw conclusions about whether that blob of phlegm truly is a cold germ carried by a former Beatle and whether it merits anything other than a spritz of disinfectant and an immediate toss into an incinerator.

Although a grilled cheese sandwich isn't the only object on which religious figures have appeared, it doesn't seem to make much difference to the believers. Many of these individuals have chosen to welcome interpretations at face value, as it were, no matter how tacky they might seem, and it is this obsessive inner need for their creations to be real that underlies their reasons for existence. If these items proved to be real (which they haven't), then the witnesses would have high PIs. But since most have been proven false, does that mean that the individuals' PIs are low?

"Immaculate perception" has been around for a long time. It was the nickname given to Plato's theory of knowledge. Francis Bacon referred to it as the necessity of "keeping the eye steadily fixed upon the facts of nature and so receiving the images simply as they are." What makes PI so fascinating is that altered perceptions so often manifest in this type of religious context. You may now rightly be asking if I am suggesting that we abandon all miraculous perceptions beyond our comprehension. In this chapter, we'll explore these questions and many others — including why some people see things that aren't there at all.

Pareidolia's Never-Ending Reach

Popular Catholic icons have a habit of turning up in the darndest places. I began the introduction to this book by citing the news story in which many visitors to St. Mary's Cathedral in Rathkeale, Ireland, believed they saw the silhouette of the Virgin Mary in a gnarled tree stump. Nearly four hundred other such appearances have been reported around the world in this century alone, including:

- On a pretzel that sold for $10,600
- On a fence post near the cliffs in the suburb of Coogee in Sydney, Australia; it has since aptly been named "Our Lady of the Fence Posts" and was the subject of a book of poetry by J. H. Crone.

- Floating on a garage door in Minersville, Pennsylvania
- As a pair of eyes on a bathroom door in Connecticut, as shown (and ridiculed) on an episode of *Penn & Teller: Bullshit!*
- Faded in a window of Mercy Medical Center in Springfield, Massachusetts

The Virgin Mary does get around (in the most immaculate of ways), but so does her only child, who has made a bunch of public cameos, including a now infamous appearance in some bathroom molding ("Shower Jesus"). Jesus even found his way to the posterior of a blissfully unaware three-year-old terrier, which its owners wisely opted not to auction off on eBay.

Although devotees herald the blessings — and, in some cases, the extra cash — bestowed on them by these holy apparitions, science has taken a more sober view, ascribing this phenomenon to a condition known as *pareidolia* (pronounced *pear-eye-DOH-lee-uh*). Pareidolia is a misperception warping our Perceptual Intelligence in which the brain interprets a vague or obscure stimulus as something familiar, clear, and distinct. Our minds misinterpret what we are seeing based on a range of factors, including our childhood upbringing, serendipity, and deeply embedded quirks in our neural processing. Sometimes the amalgam of images is so powerful that these images lace together, causing this unusual and convincing perception.

Pareidolia covers a lot of turf. Some purport that it explains a range of mysterious phenomenon, including UFO sightings (though I believe some of these are more attributable to dark lucid dreams, as described in chapter 3); messages on records played in reverse (such as the "Paul is dead" hoax, referring to how Paul McCartney died in the mid-1960s and was replaced by an equally talented impostor; "clues" can be found on some album covers and in various Beatles' songs when they're played backward); Elvis appearances (i.e., in shopping malls and gas stations); Nessie encounters (which may now have been explained as a large catfish); and as religious figures in the form of ghostly apparitions and outlines on unusual objects.

Pareidolia is by no means the sole dominion of the slightly unhinged or the religiously obsessed. Most of us have had any number of pareidolic experiences — seeing faces in clouds, puddles, rainbows, distorted developed photographs, and even on the surface of the moon. These are freaky, to be sure, but after some thought, most of us disregard them, attributing them to science or coincidental flukes of nature. Our investigation concerns those images that touch us most deeply, such as the grilled cheese sandwich, because our misguided PIs have led us to become devoutly positive that something beyond our understanding of reality is taking place. Why else would thousands of people believe that the image on the sandwich must be attributable to divine intervention?

The Faces Have It

Let's take a minute to discuss why faces are so often implicated in cases of pareidolia. We're exposed to so many daily stimuli it's inevitable that some will bear a degree of similarity to familiar visual patterns. It's not surprising that faces are so often involved, since faces tell us a great deal. From a face we subconsciously deduce information about a person's age, gender, race, and cultural background — albeit superficially and perhaps inaccurately and/ or occasionally with some measure of prejudice. We sometimes go so far as to draw conclusions about a person's origins or where the person has recently been based on her complexion; for example, a deep tan might be interpreted by our PIs as meaning a woman hails from somewhere south, such as Florida, or a tanning bed in LA. Almost instantly, when we glance at the face of someone newly introduced, our minds determine whether that person will prove friend or foe. We take this to the limit by making subjective judgments about attractiveness, disposition, personality, mood, and other qualities.

Our fascination with faces begins at birth. Newborns have blurred vision and can focus up to a distance of only about eight to twelve inches. Not only that, but they are limited to images in black and white and shades of gray. What, then, are we most exposed to when we are babies? The distorted, disproportionate, beaming faces of doting family and friends, of course. Given our early life experiences and biological programming, it's no wonder our species relies so heavily on facial recognition and goes on to discern the human countenance in so many unexpected places. This early recognition of faces seems to be the underpinning of pareidolia.

Does Religion Distort Our Perceptions?

Since there have been so many widespread instances of Catholic figures appearing on objects, one wonders why more Jewish people — who have a rich biblical history of miracles — haven't seen their iconic prophets imposed on food, garages, and bathroom doors as well. Wouldn't it make sense for at least *someone* to have reported a likeness of Moses pressed into a matzo ball on Passover? Score: Jesus sightings — thousands. Moses sightings — zero. Why?

As discussed above, our minds may program us for pareidolic experiences that began in the crib looking at Mom and Dad, but some groups seem more susceptible than others. This is not a positive or a negative but a matter of sheer number of reported instances. Strong religious convictions, or a general belief in the paranormal, lend themselves to receptivity to seeing specific faces, such as that of the Madonna or Jesus Christ. These images have been ingrained in the minds of Catholics from the very beginning; they have been depicted in paintings, sculptures, and books and surround congregants on stained glass each time they step foot in church. The faces of the Madonna and of Jesus are instantly recognizable from human characterizations of these holy figures and in their physical poses — often contemplative, with hands clasped. The respective interpretations and significance of these faces speak volumes about the PI of the devoutly religious.

Acceptance, doubt, trust, discovery, and *imagination* — these are the hallmarks of PI. Yet all our beliefs about what's true, and all our doubts, can potentially shift suddenly as new evidence comes to light. If a skeptic were to hear the face on the grilled cheese sandwich starting to talk, for example, his interpretation of its authenticity would likely change. But doubt and critical thinking police the discovery process and rightly challenge our imaginations. They free our minds of the numerous suggestions and distortions imposed on us. If I do "discover" something, my years of scientific training have taught me to question the veracity of the experience until more concrete evidence becomes available.

Religion's most significant contribution to civilization is that it was humankind's first attempt to ascribe some order to the natural world and to provide a code of rules to abide by. At the time when people first created these belief systems, there was no alternative. Today we're no longer in the dark. I will go out on a limb and suggest that my twin daughters know more about the natural order of the world than the earliest founders of some religions. Most of us in the West no longer believe that evil spirits cause disease, that lightning targets the homes of evildoers, or that natural disasters are a form of divine retribution.

To those ardent believers who to this day maintain that the Virgin Mary has chosen a grilled cheese sandwich or a pretzel on which to make her presence known, I offer the explanation that their minds have convinced them of the authenticity of these appearances because of religious imagery and upbringing; an unrelenting urgency to prove their belief systems to be valid; and an insatiable need for religion to explain the order of things at the alternative risk of perceived chaos or a world with no meaning. For many of these individuals, refuting their Virgin Mary claims would be considered offensive and blasphemous because their faulty PI has turned off their critical thinking buttons. Until science has unequivocally disproved that it is, in fact, the Virgin

Mary on that toasted bread, who are we to burst their bubbles? The truth is that all the hard evidence in the world still would not convince these believers. PI has the potential to create faith, and a force of will, that are stronger than the Great Wall of China.

Was Walter Mitty's Secret Life All a Hallucination?

Whereas having a low degree of Perceptual Intelligence might convince some of us that the Virgin Mary's face is indeed etched into a grilled cheese sandwich, the ability to recognize these images as false is what defines high PI. Take the example of the little-known but surprisingly common disorder called Charles Bonnet Syndrome (CBS) — a condition characterized by complete or partial blindness in which people see vivid, intensely realistic images of nonexistent objects — which I mentioned in the introduction. Patients with this disorder usually have damage somewhere in their visual pathway — in the eye or in the brain — causing them to be partially or completely blind.

For those stricken with CBS, the world is a place adorned with vivid yet unreal images that can linger as briefly as several seconds or last as long as several hours, appearing and then vanishing abruptly. They may consist of commonplace items (bottles or hats) or brain-bending nonsense (dancing children with giant flowers as heads). Hallucinations of crystal-clear complex patterns, people, faces, buildings, cartoons, children, and animals — often in remarkable detail — would be understandably disturbing and frightening. But these folks aren't crazy; people with CBS are often fully oriented and intelligent. The images never interact with the person. That is the one distinction between CBS and psychiatric disorders (such as schizophrenia), in which hallucinations often interact with the subject. People with CBS have generally high PI because they recognize the hallucinations are not real, whereas those with schizophrenia (through no fault of their own) lack such insight, have low PI in this area, and succumb to their engaging visions.

James Thurber, one of America's greatest humorists and author of the classic short story "The Secret Life of Walter Mitty," has been suspected of suffering from CBS. Thurber, who lost an eye at age six after his brother accidentally shot him with an arrow, had it rough. Before long, his remaining eye started failing. By the time he was forty, Thurber's life had become a blur. Instead of his world being dark and dreary, Thurber's environment morphed into a fantastic world of hallucinations and surrealistic images. Ordinary objects and experiences underwent wild transformations. He recalled scaring a woman off a city bus after mistaking her purse for a live chicken. Flecks of dirt on his windshield morphed into uniformed soldiers and crippled, apple-shaped women. Once, after his glasses shattered, Thurber wrote, "I saw a Cuban flag flying over a national bank, I saw a gay old lady with a gray parasol walk right through the side of a truck, [and] I saw a cat roll across a street in a small striped barrel. I saw bridges rise lazily into the air, like balloons."

CBS is a sophisticated type of *filling-in* response to visual deprivation. Thurber's hallucinations were phantoms interpreted in fanciful ways, caused by his eyesight miscommunicating with his mind. As with pareidolia, these kinds of hallucinations are an example of the puzzling array of phenomena that can emerge from the complex relationship between the eyes and the brain.

Walter Mitty, Thurber's most famous character, has a fertile imagination. His fantasies stretch far and wide, as he alternately imagines himself as a heroic pilot, a pioneering surgeon, and even a man who sneers at a firing squad. Only reluctantly does he return to the real world, where he is an ordinary man living a bland life. It's easy to see why he'd rather linger in his daydreams.

Thurber's hallucinations reflect a somewhat different internal contradiction. Visually they appeared real to him, but his PI aided him in recognizing and dismissing them as false. Knowing how we use our minds to see or even unsee things at will is the next step to tapping into something wonderful, imaginative, and new.

Yes, It's Fine to See Mary in Your Sandwich

So what are we to make of this habit of seeing things that aren't really there? Are we crazy if we see imaginary patterns or ascribe meaning where none exists? Is picking the face of Mary out of skillet burns or finding zoo animals in the clouds a sign of mental illness?

Like our choice in food, partners, and so many other things that are important to daily life, it seems we are prewired from birth to detect patterns and ascribe meaning. As discussed, within the first few minutes of life, most babies will focus on something that has the general features of a face. The brain's ability to think and assess quickly is a life-preserving evolutionary adaptation. Our brains are forever sorting through a vast mosaic of random lines, shapes, surfaces, and colors. The aptitude of our PI is dependent on our ability to understand the difference between what is grounded in reality and what is false. Our survival depends on making sense of these images and assigning meaning to them. This may take the form of matching them to something stored in long-term knowledge. But sometimes things that are slightly ambiguous get matched up with things we can name more easily, such as a familiar face.

Immaculate perception may be a product of expectations. Seeing the image of the Virgin Mary on a piece of bread says something about how we interpret the world, based on our hopes and dreams. We tend to project our biases onto what we see, and that has a lot to do with how we are raised, what we were taught to believe, and what we choose to accept as real to fill in the gaps and make sense of the world, give it order, and reaffirm our faith.

The Virgin Mary grilled cheese, like so many other perceived religious sightings, is no doubt attributable to Perceptual Intelligence at work among those who believe. To put it bluntly, it's just a plain old grilled cheese sandwich in scientific reality. But guess what? It's okay. We *want* to see imaginary faces and believe the impossible is possible. That's why so many of us are fascinated with superheroes and why Hollywood can't churn out enough movies with them — including the obscure ones like the *Guardians of the Galaxy* and *Ant-Man* — to quench our demand. Without our tenacious imaginations we wouldn't be unique and we would fail to question, explore, and create. Our skeptical PI would take over *too much*, and we wouldn't have the James Thurbers of the world — or the likes of Pablo Picasso, Georgia O'Keeffc, Mahatma Gandhi, Martin Luther King Jr., the Beatles, Aretha Franklin, Steve Jobs, Madonna (the singer, not the grilled cheese matron), Steven Spielberg, Gene Roddenberry, Oprah Winfrey, and scores of other influential people with brilliant, single-minded visions.

Carl Sagan said it best: "Imagination will often carry us to worlds that never were. But without it we go nowhere."

We've completed our exploration of how our perceptions can coerce us into seeing and believing some pretty outrageous things in food items and other objects, inflating their value out of proportion. On our next stop we will look at how guilt is used to manipulate our perceptions in everyday exchanges, especially when it comes to making important purchasing decisions, such as buying a car.

8

The Spell of the Sensuous

How Reciprocity Hijacks Our Perceptual Intelligence

The Roman philosopher Cicero said, "Gratitude is not only the greatest of virtues, but the parent of all others."

Imagine it's your birthday. You receive a fancy gift basket from an acquaintance you barely know. She's "a friend of a friend," and for that reason alone the two of you became casual Facebook friends. You bumped into her at Starbucks a year or so ago, and the two of you sat down for a quick chat. Although the conversation was okay, you don't care for her enough to pursue the friendship any further than the occasional click of the "Like" button on a few of her Facebook posts. Frankly, you feel you could take her or leave her. But now you're staring an expensive gift from her in the face. Do you return the favor with an expensive present when her birthday comes around — which happens to be in a couple of weeks (as you noticed on Facebook) — or will a note on her Facebook wall suffice? Do you even have to do that? Now that you think about it, you've probably dissed dozens of people by not posting reciprocal Facebook birthday wishes on their walls: oh no, what a major social media faux pas!

When someone sends you a gift, you immediately feel obligated to return one, since your Perceptual Intelligence is under assault. The *reciprocation principle*, a term coined by Robert Cialdini in his bestselling book *Influence*, is used by marketers and salespeople on a daily basis to convince us to make purchases. Car dealerships are notorious for their reciprocity games. Do you really think the salesman pours you unlimited coffee refills because he feels bad that you look so sleepy? And why do you suppose he keeps offering you free pastries, donuts, cookies, or popcorn? He's hoping you'll feel so well pampered that you'll feel you *owe* him something in return. He figures he's been so generous that you'll stick around for a while and eventually cave in, forking over a down payment on an expensive automobile instead of hunting for a better price elsewhere. Wow, that was an expensive coffee! Without even realizing it, your sense of obligation has hijacked your PI, which convinced you that this was the place to make your purchase simply because the salesman has given you stuff at no cost. How can you leave without returning the favor?

Though some reciprocal obligations are straightforward, many more of these exchanges are unequal. The total cost of everything you consumed at the car dealership in no way compares with the price of the car you are being sweet-talked into purchasing. Reciprocity has the power to alter our perceptions and influence our behavior. Your stop at the car dealership may have initially been intended as a first visit just for browsing; you had no intention of buying a car without further exploration and research. Your reality shifted from "I'll look at all the options, shop around, save up, and maybe buy the car at the end of the year when the prices come down" to ending up in a compromised fantasy mind-set. Even after having partaken of all the free coffee and snacks, you are in no way obligated to buy a car, yet you feel pressured to do so. "Why should another salesperson somewhere else get the commission when this fellow worked so hard and was so generous?" you rationalize to yourself.

Indeed, the unconscious need to go "tit for tat" often triggers an obligation to repay a perceived debt — whether positive or negative, the latter being in the case of revenge (discussed below). Living under the yoke of an unwanted or unintended debt can cripple our PIs and leave us unsure, easily exploited or influenced, and without a clear sense of the road ahead. Under these circumstances, one wonders if we can ever resist the urge to reciprocate.

Are We Programmed to Feel Guilty?

Those of us who grew up in Jewish or Italian families certainly have carried our fair share of guilt. ("It's been a week since you called," the familiar voice preaches to you. "Did you forget about your poor old mother? A call once in a while might be nice. You never know when it might be our last.") After I got engaged to my wife, Selina, my mom, as if pining for a lost love, said with a heavy sigh about my ex-girlfriend, "Oh, well, there's goes Sophie…" While most of our proverbial crosses to bear don't compare to what Jesus carried on the road to Calvary — my daughter's, for example, is cleaning up her room once a week — we put up with a lot, or at least that is our perception. But regardless of your cultural heritage, a modicum of guilt is always involved in our daily interactions. Are we born or conditioned that way? Is there ever an escape from allowing such guilt to overtake our PI and rule our lives?

Let's go back to the car dealership situation. Suppose the salesman showed you four vehicles and let you test-drive two of them. Along the way, he fueled you up with a couple of croissants and a bag of popcorn. While seated in his office, you caught sight of some pictures of his three adorable young kids, and he regaled you with stories of their soccer games and dance recitals. An hour and a half has whizzed by, and now you're filled with odd sensations: you've been there an hour longer than you expected and are running late for an appointment, you're so stuffed you could skip lunch, and you think you know this nice man and his family as well as you do some of your friends. You remember that your original intent was *just to look at cars,* and you tell yourself you really must leave — but how could you skip out after all this salesman has invested in you? Isn't his time valuable as well? Maybe he could have sold cars to three other people while you were wolfing down those croissants. Perhaps his boss watches you exit without making a purchase — what then? Would his job be at risk — and then what happens to those kids in the pictures? Will they end up with no more dance recitals, perhaps even homeless? The salesman displays a few knowing glances, makes several mentions of a "good deal," "financing with easy monthly payment," and throws in a reference to "easily drawing up the paperwork," and then...

Bam! You've taken the bait and are flopping around on the hook. Did this complete reversal from reality to fantasy (the fantasy that his kids will end up homeless if you don't buy, for example) occur as a result of your upbringing, or was it lurking in your DNA all along? We may indeed be biologically programmed to be altruistic and receptive to reciprocity up to a point, but I believe that how we react in these situations has more to do with our PI drawing conclusions based on prior experiences and on our perceptions in the moment.

Ever since the nineteenth century, when French philosopher Auguste Comte* first proposed the concept of altruism or selflessness, psychologists have debated whether people are born into the world programmed to be nice to others. A pair of Stanford psychologists recently conducted experiments indicating that altruism has other triggers and is not something we are simply born with.

A 2006 study involving toddlers found that eighteen-month-olds were willing to provide a helping hand to the experimenters without being prompted. This expression of altruistic behavior in young children aligns with what many scientists believe to be an expression of innate altruism, and the findings have served as the foundation for subsequent studies.

However, as with most experiments involving toddlers, the researchers behind this study thought there might be more there

---

* The cheese Comté is not the namesake of this philosopher, as the cheese has an *aigu* accent over the *e* in contrast to the philosopher, whose name is bereft of an *aigu*.

than meets the eye. They realized that, prior to the testing, they had engaged in a few minutes of play to ensure that the toddlers would feel comfortable with new people in an unfamiliar setting. This brief interaction might have inclined the toddlers toward altruistic behavior, influencing the outcome of the experiment.

According to the researchers, all human beings — children, in particular — seek social cues, selflessness being a prominent one. Playing with a child lets her know that another person will care for or be kind to her. "These actions communicate a mutuality, and the child responds in kind," the researchers concluded. During the pretest play period, the kids who engaged in *reciprocal play* were three times more likely to engage in altruistic behavior than those who engaged only in *parallel play* (no involvement with others), suggesting that altruistic behavior may be governed more by established relationships and environment than by biology.

As adults we have learned from hard life lessons that compromise and concessions are frequently the only things that can help win over allies, settle a debate, or bring closure to a heated dispute. On the extreme side, many businesspeople ("backstabbers") and politicians ("corrupt," "dirty") are seasoned pros at using reciprocity to hold an advantage in negotiating, gaining supporters, or winning a case for a client. They easily rationalize and justify dubious behaviors ("it's all business," "all's fair in love and war," and "all politicians do it") shoving their victims' PIs into the kind of compromise mode that would make using kryptonite on Superman seem like child's play. We've all heard the clichéd expressions "I'll scratch your back if you scratch mine" and "One hand washes the other" and immediately associate them with the underhanded politician who will support a senator's proposal if she backs his. (I was so gullible all those years when I thought the political term *pork barrel* referred to the amount of barbecued ribs eaten by a politician the night before a big election.) Or perhaps the opposite of "you help me, I'll help you" occurs; she didn't support his proposal last time, so there's no way he'll be on board for hers. In cases like these, whether in politics, business, or other areas, we determine whether we must reciprocate out of guilt, fear of retaliation, or the need to have an ally in our pocket for later on ("Remember the time I supported you on that proposal? Now my proposal is coming up on the agenda...").

In his book *Influence*, Cialdini cites what I believe to be another PI killer: *concessions*. These occur when you're misled into believing one option is better than another, even if it's not. Concessions trigger a reciprocation response because they involve something known as *perceptual contrast*. Suppose a client owes you $10,000 for services rendered, which you diligently performed in fine form. Despite your having fired off a number of overdue invoices and increasingly hostile letters and emails, a year goes by without any acknowledgment from the client. At your wit's end, you sic your attorney on the client, with a warning letter: either he pays up or you're going to court. Finally, the client gives you a call and offers a halfhearted apology: "I'm sorry about all this, you know me — I'm not like that. I always pay my bills. Look, nobody wants to go to court. Think of all the time and wasted money on lawyers. How about we just split the difference and call it even. Fair?"

No way — it's not fair at all! You've waited a year for your payment, spent time chasing after him, paid a lawyer to write the threatening letter — and now you're going to accept *half*? Unfortunately, while in the moment your mind doesn't see it that way. After a year of frustration, you've been manipulated into thinking that settling for less is somehow fair. The client has a point in that you don't want to go to court, since it was all a bluff anyway; not only would it be expensive, but what if you were to lose the case? In this proposed scenario, the client is offering you *something* when the alternatives in your head make it appear as if you won't get a dime, or worse, as if you'll lose even more money on expenses.

No one is completely immune to being on the receiving end of reciprocal obligations and having his or her PI reduced to ashes. On many occasions I have felt the powerful pull on my own reciprocal instincts, just like everyone else — especially in challenging social situations.

The Even Balance Sheet of Gift Giving

You've no doubt experienced this at some point in your life: you're preparing to attend the wedding of a cousin twice removed, whom you've seen on perhaps only four occasions in your lifetime. You purchased a suitable wedding card and are writing out a check to slide inside it. You deliberate for several minutes about how much you should give and can't seem to come up with a number that makes sense. Do you write the check out in an amount that factors in your two kids attending the ceremony and party? Do you weigh in how fancy and elaborate the party will be?

Unsure of the etiquette and the going rates for wedding gifts to distant cousins you barely know, it hits you like a flash: *What did this cousin give you for your wedding?* It was eight years earlier, she wasn't married then and didn't have kids, but at least there is some logic to drudging up this obscure fact in an attempt to find a comparison to serve as a measuring stick. You ask your spouse if he recalls the amount and he says, "I remember it was cheap — like twenty-five bucks or something like that." With no other direction to go in, you multiply the twenty-five by four to factor in all your family members and scribble out a check for $100. Mission accomplished!

Now, suppose the cousin had been flat broke at the time and in debt, having just lost a job and struggling to pay off college loans; the $25 may have been all the cousin could afford at the time. Would that have changed the amount you wrote on the check?

But what if your spouse had incorrectly recalled the amount of the gift and your cousin had actually given you a lot more than that? Or maybe you forgot that in addition to the check she had also purchased a prized bread maker for you at the bridal registry, making the combination quite a generous gift indeed. Do you deduct points for bad gifts like a cheesy, outdated wine holder?

Let's consider an entirely different scenario. The spouse remembers that the cousin had honored you both with a *$500* check eight years earlier. "Oh, my goodness," you think to yourself, with the pen shaking in your hand. "How can we possibly afford that? Why did my cousin have to give us *so much*?"

I happen to know an individual whose close cousin didn't give him a wedding gift. No fewer than *ten years* went by, but it didn't lessen the lingering disappointment that this beloved cousin had done the unthinkable and stiffed him. One day a mysterious letter appeared in my acquaintance's mailbox. Inside was a check for $100 from the cousin, dated ten years earlier; an envelope with various crossed-out addresses and markings indicating the recipient "moved"; a crumpled wedding card signed by the cousin; and a handwritten letter from a total stranger. The gist of the letter was something like this: "You don't know me, but this gift has been floating around in the mail to incorrect addresses for such a long time, I decided to look you up on the Internet myself. You moved around a lot, but I finally found your current address (I hope!) and decided to send the card and check along. Congratulations on your ten-year anniversary!"

On receiving this miraculous long-lost gift, my acquaintance immediately called the cousin and thanked her profusely, saying he had an amusing story to share. It didn't matter that the check was no longer valid (the account didn't exist anymore); the point was, he now felt guilty for having held an erroneous grudge and misperception of his cousin for all those years.

Yes, there will always be people who believe in "tit for tat" with gifting and who treat life as if it's a balance sheet. Will God or the clerk at the pearly gates prevent you from entering heaven if you came up short in gifting — or the reverse, shut you out because you were showboating with the inflated size of your gift checks? Reciprocity with gift giving is a no-win situation, and there is no reason to lower your PI and allow this kind of mindset to drive you crazy. A caring friend or relative will appreciate that you joined in the celebration and were thoughtful enough to provide a gift. Do not let other family members or friends influence your PI either by trying to compare amounts; it will only make you feel guilty or cheap (for giving too little) or stupid and reckless (for overspending).

Suffice it to say, gifts are a wonderful thing, and you needn't stress out about them or overthink things. The key is to avoid reciprocation wherever possible. Always give what your heart tells you, based on what you can afford — no more and no less.

An Eye for an Eye: Reciprocal Bad Behavior

Revenge is a dish best served cold...or is revenge *sweet*? These phrases have become clichés because to many people the thought of "getting even" with someone provides immediate PI gratification. In fact, just picturing the steps involved in conceiving the revenge and exacting it are enough to send a blizzard of sparks to the brain, simulating something intensely pleasurable, like eating a delicious, velvety chocolate cake without a care for all its calories. Exacting revenge tends to be calorie-free — although you might end up in jail.

Our brains are hardwired to enjoy revenge. MRI studies have shown that when we merely think about revenge, dopamine floods the pleasure areas of our brains — the same reaction as when we eat that piece of chocolate cake (dopamine release is the reinforcement mechanism behind those with chemical or drug addiction). In all likelihood, this is an evolutionary leftover from when primitive human beings had instincts informing their brains that revenge was essential for survival. If a tiger attacked a band of prehistoric people, killing some and maiming others, the leader would motivate his surviving followers to gather spears and rocks and follow in pursuit. The revenge motive appeals to the group as a rallying cry because deep down they know the tiger might think they are weak and will someday return to draw more blood if they fail to make a show of strength. The collective grunts as the tribe runs in support of the leader acknowledge that everyone shares in the revenge and is on board with the plan, giving them a necessary advantage: strength in numbers *and* a good cardio workout.

We'd like to think we've evolved since then, but some people might say we haven't come far enough. There is plenty of reason for humans today to want to display reciprocal bad behavior. Think about all the times you wanted to get back at someone in the office who stole your idea or got a promotion you felt you deserved; think of when you were a child and your sibling ratted you out for taking your parents' car on a joyride and putting a dent in the fender.

Those of us who grew up with siblings know firsthand what revenge feels like (on both the giving and receiving end). When we are children, our PIs are low because they haven't matured enough to respond properly to perceived attacks. If a brother intentionally spits in the face of his older sibling, we can rest assured that "it's on" and blows are imminent. In my house, my darling daughters become not so darling when one thinks she has been wronged by her sister, even if by total accident. Once I watched one of my daughters inadvertently step on the other's foot. My wounded daughter didn't give it a second's thought; she curled up her fist and took a retaliatory swing — *pow*! (Truth be told, sometimes I enjoy having ringside seats before I intervene, just like an NHL hockey ref.)

Is There Truth to the Adage "A Woman Scorned..."?

The topic of the difference between male and female brains can be fairly tricky to navigate.* With that in mind I'll try to tread lightly and stick to the science about what we know of the brain and how men and women perceive revenge differently. Put simply, men have a lot of testosterone, whereas women have a lot of

---

* I think I'm pretty fair when discussing female and male behaviors. After all, I added my wife's last name (Boxer) to my last name (Wachler) when we were married, which also makes me one of the few men in the country with a maiden name. John Lennon tried to change his last name to Ono Lennon, but British law precluded last name (surname) changes, so he had to settle for changing his middle name to Winston Ono.

estrogen; this basic variance in biology causes men and women to react differently to feeling wronged.

When a man is threatened, his testosterone levels reduce fear, which allows his aggression to be unleashed. In one scientific study, when women were injected with testosterone, their brains displayed charged-up activity in the amygdala, and the women responded far more aggressively to situations. Now, if women generally have less testosterone and are thus genetically less prone than men to hostile, over-the-top responses, why has so much attention been given to the notion that women may be *even more vicious* than men when mistreated and far more likely to dish out payback?

The origin of this perception goes back to 1697 and *The Mourning Bride*, a British play by William Congreve. The full quote goes like this:

> Heav'n has no Rage, like Love to Hatred turn'd,
> Nor Hell a Fury, like a Woman scorn'd.

Back in the early 1990s, a scandal took place in Manassas, Virginia, that captured the attention of gossipmongers everywhere: the tale of Lorena and John (Wayne) Bobbitt. Remember that infamous couple? As the story goes, Lorena (the woman scorned) cut off her husband's penis while he was asleep as punishment for his infidelity. *Ouch* — the terrifying female psychos portrayed in the suspense films *Play Misty for Me* and *Fatal Attraction* have nothing on Lorena Bobbitt!

When a woman is said to fall "head over heels in love" with someone, dopamine — yes, the very same neurotransmitter I mentioned earlier as being involved in pleasurable revenge — floods her mind. Studies have shown that this is an intense, potent response (think of it as a love potion in the brain); if a woman is knocked out of this blissful feeling, such as by an unfaithful partner, her dopamine nose-dives, and anything redeemable about him is tossed out the window. In extreme cases you'll hear of jilted women who key their lovers' cars, have sex with their lovers' brothers or best friends, cause embarrassing scenes in malls, or post all the sordid details on the Internet and on highway billboards. And that's just on day one of war!

That is not in any way to say that men are saints. Men are wholly capable of doing some pretty terrible, abusive things when their lovers dump or cheat on them. (Here I want to be clear that I am not referring to those who are verbally, emotionally, or physically abusive or just disturbed stalkers; these men can be triggered by nothing at all before going ballistic.) These passionate reactions tend to be in the heat of the moment and may involve immediate threats, breakage of objects, and physical abuse; some ultra-extreme examples have been known to lead to suicide and murder (think of the O.J. Simpson case). The difference is that, aside from psychopaths (who will act in all kinds of dangerous, unpredictable manners if unmedicated), men tend to have immediate hostility and aggression as long as the testosterone and endorphins are clouding their heads; once they're released and out of their systems, these men usually simmer down after a while, realize they overreacted, feel remorse, profusely apologize with heads lowered, and order flowers. *A lot of flowers.* Some women, on the other hand, have been known to hold longer grudges when jilted and often have flare-up feelings of revenge as the storm of dopamine comes and goes.

In any case, if you are ever feeling jealous rage — male or female — try to concentrate on pleasant memories of the individual in order to readjust and refocus. Reciprocal bad behavior is a sign that your PI fell off the cliff; if this controls your impulses, it can lead to your inflicting harm on others and possibly even serving jail time.

In the next chapter we'll investigate the impact of real celebrities (unlike the John Bobbitts of the world, who become famous solely for regrettable reasons) on our Perceptual Intelligence.

9 Star Time

Blinded by the Glare of Celebrity

You may recall that in chapter 6 I discussed how some larger-than-life sports figures rapidly ascend to glory but then plummet to the ground with a *poof* of smoke that would do justice to the image of Wile E. Coyote landing at the bottom of a cliff. I also described the fan fervor surrounding sports teams that can distort Perceptual Intelligence and overtake fans' lives to a shocking degree. Celebrities — by this I mean anyone in the spotlight, including film/stage/TV actors, rock stars, comedians, Instagram icons, YouTube famers, and reality TV stars with more than fifteen minutes of fame — can have an even more mesmerizing influence on our PIs, causing us to feel that anyone famous is magnetic, especially in terms of success, money, talent, and looks. Many people are intimidated by celebrities and/or long to *become them* because their minds have been manipulated by the image and lifestyle and fooled by the makeup and surroundings.

The influence of celebrity on individuals and on our culture has become so prevalent, due to the immediacy of social media, that our minds are coaxed into believing that famous people are superior to the rest of us; this has led to a phenomenon known as the "halo effect." Often people take the words and images of celebrities (and politicians) more seriously than those of scientists, educators, or thought leaders who are genuine experts on the subjects, which leads to ignorance and battered PI. Even the pervasive social media word *follower* has connotations of being led by a Pied Piper down a certain path. In the modern era, it has become standard for famous individuals to make statements on everything from politics to the environment to human rights to child vaccinations — whether or not they understand the issues involved (admittedly, some are more knowledgeable than others).

Whether or not we realize it, every one of us has fallen victim to the halo effect at one time or another. In this chapter we'll tackle a few of the ways the halo effect manipulates our thoughts, beliefs, and self-images — from the world of advertising to social media.

Halo, Good-Bye

The halo effect, as it's usually described, is a type of perceptual bias in which our overall impression of a person influences how we feel and think about her or his character — even if that impression isn't based on fact. No one is completely immune. From the early days of radio and television to the Internet of today, celebrities have had such significant influence on fans and followers that many of them have become as well known for their product pitches as for their professional accomplishments. Years ago Hall of Fame baseball player Joe DiMaggio became forever linked with "Mr. Coffee." More recently, William Shatner (a.k.a. *Star Trek*'s Captain Kirk) brought attention to Priceline, doing entertaining martial arts moves in TV commercials. Even actor Wilford Brimley has gotten in on the act, channeling his stodgy-old-man persona to appeal to viewers of a certain age watching commercials for Liberty Mutual and Quaker Oats. And let's not forget the timeless ads from female stars: Brooke Shields (Calvin Klein), June Allyson (Depend underwear), and Martha Raye (Polident denture cleanser).

When trusted high-profile celebrities appear in commercials, pixie dust is blown into the minds of viewers as they magically associate celeb auras with the products being pitched, making the items seem far more credible and desirable. Our PI can be dramatically lowered by the right spokesperson choices, as we'll discuss below. If Brooke Shields did karate chops for Priceline or Wilford Brimley advertised Depend underwear, would they have had the same impact?

In this day and age, things have "evolved." Now celebrities can parlay their halo effect to influence us in unusual ways. In the past, TV personality Jenny McCarthy was outspoken against vaccines, stating that the risk of autism is more significant than that of measles and chicken pox. In social media, the rich and famous are influencers on a monumental scale, with the ability to plant a product seed in the minds of millions of followers with just a 140-character tweet. Though these exchanges seem harmless on the surface, there are hazards since it may not be at all clear whether these were paid ads, and they tend to occur at the speed of light.

Kim Kardashian West receives anywhere from three quarters of a million to a million dollars per gig to be a product spokesperson and receives $10,000 per tweet or Instagram post. Why are celebs paid so much? Because the advertisers know that often whatever celebs such as Kim Kardashian West tweet or say will catch the attention of millions of their followers as a result of the halo effect. A number of followers simply don't care whether the claims are true and believe them at face value; Kim's typed characters are more than convincing enough. (More at the end of the chapter about Kardashian West.)

The halo phenomenon isn't limited to celebrities or even just perceptions of people. Teachers fall prey to the halo effect when evaluating certain students they think are a grade above the others. A boss may be drawn to a single attribute, such as an employee's enthusiasm or attractiveness, which then inadvertently colors the evaluation process. Consumers who purchase food labeled organic perceive it as tastier, less fatty, lower in calories, higher in fiber, and more nutritious than the nonorganic choice, even if it isn't.

Awareness of the halo effect doesn't make it any easier to avoid its influence on our PI and our resulting decisions, especially when we view someone, something, or a situation as black-and-white and are satisfied with this perspective, locking it into place. Both the gullible and the cynical can fall prey to bias when it comes to the halo effect, especially with regard to marketing, publicity, and advertising.

An extreme and far more sinister result of the halo effect is celebrity stalking. These individuals who become stalkers have become so consumed by thoughts of meeting *or becoming* their heroes and role models that they trail after them wherever they go — often popping up unannounced in both public and private places for more than just an autograph. It's become something of a rite of passage for superstars to adopt a stalker or two along with their hordes of giddy groupies, fans, and followers. Beyoncé, Selena Gomez, Jessica Simpson, and Jennifer Lopez are just a few celebrities who have been threatened by stalkers. This phenomenon is by no means relegated to women; actors Alec Baldwin, Colin Farrell, and John Cusack have all had moments of terror facing deranged fans. And let's not forget the attacks on two Beatles: John Lennon, who was murdered in front of his New York City home in 1980, and George Harrison, who was stabbed inside his home by an intruder in 1999, though he survived the attack.

Clearly, it can be dangerous being famous — so much so that a friend of mine who works for a well-known talent agency in Los Angeles informed me that her company created a "crazy line," as they refer to it, for callers who demanded to speak to celebrities. (They are put on hold, transferred to another line, and presented with unbearable muzak until they get tired and hang up.)

In Martin Scorsese's 1982 film, *The King of Comedy,* Rupert Pupkin (Robert De Niro) plays a wannabe comic who is so in awe of his comic idol, Jerry Langford (Jerry Lewis), that he kidnaps him and creates his own phony talk show set in his home (with cardboard cutouts). "A lot of you are probably wondering why Jerry couldn't make it this evening. Well, he's tied up, and I'm the one who tied him," Rupert says on camera. "You think I'm joking, but that's the only way I could break into show business — by hijacking Jerry Langford."

It's perfectly fine to admire talent, looks, and charisma and even be inspired to try your hand in the entertainment business. But once a person becomes so enraptured by a celebrity that he or she doesn't have a life — or worse, becomes a Rupert Pupkin — then Perceptual Intelligence sinks, sometimes to the point of no return.

Since we've already ventured into the world of humor with *The King of Comedy*, let's now take a look at how our humor plays a role in our PI.

Are Comedians "Dirty"? It's in the PI of the Beholder

Few things are more fragile and subject to interpretation and personal taste than comedy. What one person finds offensive is hilarious to another. Woe to the low-PI comedian who pokes fun at Israel at a comedy night performance at a synagogue in Long Island, or the low-PI white comic who uses the N word while doing standup at the Apollo Theater in Harlem. Politics, religion, sex, race, disabilities, and death can be excellent subjects for comedy, and comedians must take daring risks to build their careers, reputation, and art. But comedians must also exercise caution: our PI necessitates that the timing and the audience must be 100 percent right and that the stage image created by the performer befits the material. Only a few comedians can get away with stereotypical humor to general and ethnic audiences — Don Rickles comes to mind — and that is because they have created stage personas so specific that audiences are in on the joke and know the comedians are "just kidding."

Going back in time for a moment, Lenny Bruce, the comedian and satirist who revolutionized modern standup comedy, had what may be the mother of all mother-in-law jokes (I may be slightly paraphrasing): "My mother-in-law ruined my marriage. One day my wife came home and found us in bed together."

Is this offensive? Well, perhaps not by today's standards. The joke and our evolved reactions to it illustrate how culture changes our PI over time. Perceptions of reality (and morality) at that moment dictate what constitutes obscene and inappropriate humor and what is benign in another era. In the 1960s, Lenny Bruce's mother-in-law joke crossed the red line of what was appropriate to many people. Today, social norms have shifted that red line way back, and, as a result, our PI will accept far raunchier humor.

In his day, Lenny Bruce was not only prohibited from practicing his art in many clubs because of his use of profanity and focus on taboo subjects (like fornication with a relative), but he was *arrested* on several occasions. The years of court battles caused significant anguish to Bruce, who saw himself as much more than a comic spewing jokes. Many famous comedians — including Henny Youngman, Milton Berle, and Jackie Gleason — told scores of mother-in-law jokes but never went as far as Bruce did. Bruce was ages ahead of his time challenging social mores and set the stage for Robin Williams, Louis C.K., George Carlin, Richard Pryor, Eddie Murphy, Amy Schumer, Sarah Silverman, Margaret Cho, and so on. These and many other comedians tend to have unusually high PIs because they see and reveal truths (exaggerated for effect) that audiences may have missed (or never expressed aloud). Comedians, whether or not they are "dirty," help us laugh at ourselves and our foibles, widening our perceptions and improving our PIs — usually to expose the absurdities and falsehoods of our daily lives.

LAUGHING ALL THE WAY TO THE BEDROOM

Even comedians at the bottom of the pecking order have observed the effect of the halo's glint of celebrity on the

PI of comedy club audiences. Being onstage in a small comedy club bestows an aura of celebrity on those in the spotlight. After the show, comedians often hang around the bar, where chances are high that adoring fans will flitter to them like moths to a flame.

The endearing industry term for these comedy groupies is *chuckle fuckers*. Comedians must have some measure of confidence, charisma, and command to hold their own onstage, all of which are considered seductive traits. Mainly, the fans are drawn in by the hint of celebrity status (granted, sometimes *small-time* celebrity) and by the fact that these performers are more attainable conquests than rock stars or hunky actors.

Reality TV and Selfie-Gratification

For years tabloids made a name for themselves by exaggerating (and often fabricating) celebrity addictions, infidelities, divorces, love triangles, cosmetic surgeries, meltdowns, weight gains, and alien connections. The halo effect of celebrities detailed earlier in this chapter included the overall mystique surrounding stars; the celebs reported on were so far above average people that even the most absurd and negative new stories — with badly shot, blurry photographs often intentionally distorted to make them look obese, harried, and unkempt — became plausible and engrossing to voracious readers. Nary a week would go by that Oprah wasn't on a revolutionary new weight-loss plan, Michael Jackson wasn't revealed as an alien, and Liz Taylor wasn't falling in love for the gazillionth time. Tabloids have by no means gone away (in fact, it's just the opposite, since there are now two national tabloids with *National* in the title), nor has public fascination with the lives of the rich and famous dwindled. People will continue to have an insatiable need to learn the latest on the never-ending saga of Jennifer, Angelina, and Brad. But something has happened over time: the gossip-hungry, celebrity-addicted universe has shifted format to include all-new sources and formats, namely reality TV and the Internet, especially social media.

Reality TV is a unique form of entertainment. Usually the shows feature good-looking people who are brought together (and sometimes egged on by producers) to whine, argue, battle, and scheme to boost ratings. Why do the boorish entanglements of Hollywood wives as they get their nails done and plan lavish dinner parties mesmerize such wide audiences? These individuals generally aren't actors, comedians, or rock stars. The fact that they're on TV dolled up (or not) and are connected with celebrity in some fashion (even crime, as in the case of shows about mob wives) make them irresistible both to younger audiences and to the prior generations who have moved on from trivialities about Oprah and John Travolta. For viewers who might be disappointed with their own lives, reality TV is an opportunity to peek through a keyhole into other worlds.

Tom Sandoval and Scheana Marie are two charismatic and endearing patients of mine (Tom had Fortified LASIK®, an advanced LASIK surgery that I invented, and Scheana has colored contact lenses) who star in the hit *Vanderpump Rules* reality show, which has captivated a growing cadre of fans. They are nice folks going about their lives, but being broadcast on TV has transformed them in the perceptions of their fans. Going much further than simply printing an article in a tabloid like the *National Enquirer*, reality TV holds a camera in front of these people in and out of their makeup and fancy clothes and with their baggage and warts exposed. Reality TV offers an opportunity to observe these celebs as they act like real people, make visible mistakes, and have emotional outbursts.

The reality TV phenomenon has seamlessly translated over to the Internet, where the celebs themselves can control what their followers see about them on Instagram and other sites, with rippling effects. Case in point: My patient Corinne Olympios of *The Bachelor* reality TV show had Fortified LASIK® with me.

Beforehand, Corinne had posted about the upcoming procedure to her Snapchat followers. The responses I received from her post were impressive. Social media stars can wield tremendous power that can be used to raise awareness of important topics. A good example of using a social media platform to help others is the endearing ACE family, whose YouTube channel (The ACE Family) has over 1.8 million subscribers. They posted a video with a family member who has Keratoconus. He was met with the emotional surprise of being flown out to see me to have Holcomb C3-R® and Intacs® treatments, which can improve vision and prevent an invasive cornea transplant. Less than twenty-four hours after this video was posted, it had received a staggering 906,633 views and more than 26,000 comments. The influence and engagement of social media on devout followers cannot be underestimated.

Meanwhile, "selfie stars" can flaunt themselves at any time of day by photographing themselves with their smartphones and posting the images on Twitter, Instagram, and Snapchat for thousands or millions to see. The selfie generation has officially taken over, and everybody is invited to the party...or at least to watch the party. Kardashian West, who has been dubbed "the queen of selfies" and "breaker of the Internet," published (and republished) *Selfish*, a book of photographic selfies. Many consumers purchased the book for their coffee tables, and probably a great deal more surreptitiously flipped the pages of the seductive shots while browsing the bookstore aisles.

Kardashian West eyes herself semi-nude in the mirror with a seductive look; the portrait is created and sent through social media from the subject directly to her followers, who can react and respond in real time. From a marketing perspective, it's brilliant in that it creates an even stronger bond to the fans than their merely reading about her in a celebrity magazine. (You'll find more on celebrities and marketing in chapter 12.) Followers rarely think that this is marketing, since their minds are blinded by the halo's glint. While not much of an angler, Kim K. is the most successful fisherman (or fisherwoman?) there ever was, boasting fifty million on her hook at the same time. The fan emotion and immediate gratification of this one-sided social exchange — which seems two-sided because of the viewer's ability to comment — goes right to the center of PI and the redefined, faux intimate connection between celebrity and fan. Kardashian West has total control of the image of herself she wishes to convey. The relationship is synergistic with her followers as she forges a bond between them and her. Her fans receive the newest selfie on a regular basis (just like from a real friend on social media), reflect on it, judge it, remark on it, share it, and absorb it into their PI fantasies.

Let us now move on to how our Perceptual Intelligence interprets our favorite three-letter word: *nap*. Oops, I mean *sex*.

10 Persexual Intelligence 101

Each to His/Her/Their Own

Before we begin our exploration into carnality with regard to Perceptual Intelligence, let me state for the record that I don't pretend to be a sexologist or sex therapist. My medical practice focuses on the eyes and vision. That said, I know that passion and sex have driven human interaction and decision making since the beginning of time, resulting in misperceptions, misunderstandings, and fallacies that reveal whether we have low or high PI in this arena.

Our fantasies, our ability to distinguish what is real from what isn't, and what sexual acts we engage in tell us a great deal about society and ourselves. In some cases, refusing to accept the natural order of things, such as the fact that masturbation is a normal biological need and not connected whatsoever to morality, goes against all available scientific evidence and demonstrates low PI; such denials of fact are along the lines of whether or not the Earth is round and global warming exists. The opposite extreme also holds true. Those who fall prey to illegal and/or immoral (child pornography, incest, rape) fantasies have low PI with dangerous implications to fellow human beings as well as to society as a whole.

When it comes to controversial areas such as pornography, masturbation, infidelity, sexual preferences, and sexual deviances, at what point do we cross over the line in our belief systems? Recognizing that we are surrounded by sex on the Internet, on TV, in the movies, in advertising, and virtually everywhere else we look in our daily lives, how can we be certain that our own perceptions haven't been unduly influenced and distorted beyond the norm? In this chapter we will explore how sexual inputs and our interpretations of them both guide and mislead our PI.

The Internet: Sexual Disneyland, or Virtual Sodom and Gomorrah?

If you're among the few human beings who have been able to resist Internet porn and judge others who have viewed it, consider the following statistics (global unless otherwise indicated):

- 30 percent of all Internet traffic is porn.
- The top porn site gets as many as 350 million unique visitors a month (which is almost the equivalent of the entire US population).
- 64 percent of American men view porn at least monthly (this percentage is nearly the same among Christian men).
- 79 percent of men in the age group eighteen to thirty view porn at least monthly.
- 67 percent of men in the age group thirty-one to forty-nine view porn at least monthly.
- 55 percent of married men view porn at least monthly.
- One out of every three women watch porn every week.

That's a lot of porn watching! Folks are so busy downloading explicit videos and pictures, it's a miracle anyone has time to earn a living, mow the lawn, go to yoga class, or watch the latest Pixar film with the family. When it comes to sexual proclivities, interests, and curiosities, pretty much anything we could possibly fantasize about is right there at our fingertips — twenty-four hours a day, seven days a week, all around the globe — free for the viewing with just a few strokes of the keyboard. Never before has the world experienced anything like this deluge of promiscuous content, although concupiscent images, dreams, and behaviors have long existed, as has the representation of erotic images. The difference is that today we have the technology to easily photograph, video, share, search, and find pretty much anything we may wish to experience.

I'm Too Sexy for My Toga

Perspectives on illicit content have changed over the course of the centuries. What we consider pornography today has been ever-present and accepted throughout history in some visual manner or form, going back to ancient Greek and Roman representations on pottery, drinking cups, lamps, jewelry, and other items that were commonplace in those times. It's always with some trepidation that we venture to an art museum with our young children and explain to them why the sculptures of men and women are almost always nudes (and typically missing an arm or a nose). The Greeks and Romans didn't see sexual acts as anything other than a daily life activity, however, and depictions, including paintings on restroom walls (as can still be seen in Ephesus), of nudity and fornication were displayed for all to see in religious temples and all kinds of public meeting places. The PI of these ancient people was high in this regard, as they saw nothing lewd about representing the human body and its natural sexual proclivities — whether hetero- or homosexual. In fact, not only do the primary remains of the baths in ancient Pompeii, Italy (wiped out by volcanic eruption in 89 CE), reveal art that we would consider sexually explicit by today's standards, but they show people using toys, such as dildos, to heighten their pleasure. Suffice it to say, the ancient Romans were quite comfortable with their sexuality — perhaps too much so when it came to orgies, sex with slaves, and so on — and they even portrayed their gods as pretty darn horny (especially Zeus, who fornicated with women in animal forms, such as bulls).

India took things a step further. Somewhere around the second century, a Hindu philosopher named Vatsyayana created the world's first known relationship book/sex manual, *The Kama Sutra*. In addition to offering sage advice on life and love, *The Kama Sutra* describes all the known sexual positions in extensive detail. Achieving sexual pleasure was considered essential to a healthy life and a fulfilled relationship, and acting out the myriad positions described in the book was believed to be a method for obtaining spiritual bliss with a partner.

Shifting Perceptions: Clampdowns in the Name of the Lord

We know that many ancient cultures, including the Greeks and Romans, were accepting if not outright encouraging of outward sexuality and visual representation of sex acts. But Perceptual Intelligence is a dynamic process, and history is also filled with countless examples of societies with equally repressed cultures. It seems the repression primarily evolved over time as people became more educated and sophisticated and religious doctrine seeped into and dictated every aspect of life.

Victorian English culture, for example, eschewed the concept of sex for pleasure, and the Church deemed pretty much everything about the act sinful, unless it was specifically done with the intent of procreation. Even birth control was banned. Ironically, clamping down on sex only served to make prostitution more rampant in London during that period. Time and time again, the sex business has boomed under the restrictive pronouncements of the Church.

Since the first Puritans settled in America, religious influence has created all kinds of misconceptions about sex, especially as related to what is considered pornography, which sexual behaviors are acceptable, and whether masturbation is harmful to mind, body, and/or soul. The Bible itself paints a rather narrow view of this act, describing how Onan was punished by God for having spilled his seed. (The term *onanism* is frequently used to refer to masturbation.) Since then, priests and preachers have lectured about the ills of masturbation and warned that those who pleasure themselves will face dire consequences. To be sure, over the years untold millions of Christian Americans have been brainwashed into thinking that if they touched themselves, one or all of the following things were sure to happen: (1) their private parts would fall off, (2) they would go blind, or (3) they would go straight to hell.

THE MYTH OF THE CHASTITY BELT

To this day, many people remain convinced that chastity belts were used during medieval times to ensure that young maidens maintained their virginity until their knights or lords returned to tie the knot in proper religious wedlock. This couldn't be further from the truth. Chastity belts didn't come into vogue until the nineteenth century, when physicians (even in the United States) prescribed them mainly to young men to prevent masturbation, which they believed was detrimental to health, potentially causing mental illness, heart disease, or cancer.* (Ironically, the opposite is true; masturbation

---

* The United States certainly has some dirty little secrets in its past regarding the intersection of sex and medicine. In the early 1900s, in California, eugenics (involuntary surgical sterilization) began to be performed on a number of sexually active girls (mandated by their parents) who were simply responding to their hormones. An enthusiastic twenty thousand sterilizations — including those on men with mental illness — were done between 1909 and 1960 in the Golden State (not so golden for these folks), which equates to one-third of the total eugenics procedures in the entire United States.

has been scientifically proven to reduce the risk of prostate cancer and is considered something of a biological need.) At about that time, chastity belts were also known to serve as protection for some female workers to protect against undue advances from their bosses and colleagues. Even so, can you imagine being the poor boy forced to wear a chastity belt under his swimming trunks when hopping into a lake to cool off with his friends on a sweltering summer day?

From the 1960s to the 1980s, Father Morton A. Hill, a Jesuit priest and one of the founders of Morality in Media, launched several campaigns against pornography and even worked with the Lyndon B. Johnson and Ronald Reagan presidential administrations to rid the world of sexual content. Father Hill was by no means alone in this crusade, and many other religious leaders and evangelists (such as Jerry Falwell) since then have made banning porn a significant part of their missions. Over the years, religious factions — especially the Christian right — have gone to even greater extremes, not only trying to ban pornography in magazines and films, but also actively censoring artists, writers, television programs, films, and music for content less offensive than what you might find exhibited in the Metropolitan Museum of Art in New York.

The PI Verdict: Smut or Sexual Aid?

The unanswerable questions continue to be asked: Who determines what constitutes pornography? And where must the line be drawn? Perceptual Intelligence plays a key role in our thinking. Religious fanatics with low PIs will go to extremes of trying to cover up breasts on an ancient Roman sculpture, ban literary classics such as *Lady Chatterley's Lover* and *Madame Bovary*, and add warning labels on CD and record sleeves of rock and rap music. On the flip side, feminist organizations battling pornography have a genuine point when it comes to images of women involved in demeaning, degrading, and/or violent acts against their will and anything remotely smacking of child pornography, among other heinous crimes.

Is the huge volume and instant accessibility of Internet porn bad for your PI? I believe that it's within PI norms to once in a while give in to the temptation, in the privacy of one's home, of looking at or watching the fantasy of choice. Anything crossing over to abuse of oneself or others (especially when underage children are involved) reveals deeper underlying psychosexual issues that likely require some form of treatment.

Strong sexual PI means that viewing and enjoying whatever floats your boat (outside the aforementioned issues, of course) is fine, as long as you don't become addicted to the point of wasting several hours of a day at the keyboard and letting this activity interfere with personal and work lives. For those with low PI, sexual fantasies become distorted perceptions, and it's possible for these individuals to cross the line and act on them — never a good thing.

Keep Your Hands to Yourself: Masturbation and PI

Our sex drives originate during puberty, when our hormones explode into an all-out frenzy, while, ironically, we have zero emotional intelligence or experience to control it. We are dependent on the sexual perspectives we formed during our upbringing (according to Freud, starting right when we are introduced to our mothers' breasts) and the people, images, and settings around us, although our proclivities and tastes are locked deep within our DNA and our brains. As renowned sex therapist Dr. Ruth Westheimer once said, "When it comes to sex, the most important six inches are the ones between the ears."

Teenage males have particularly low sexual PI while trying to process racy thoughts pretty much every minute of the day. Their lurid fantasies are voluminous and, quite simply, all consuming. The average teen boy is known to masturbate to thoughts of the latest YouPorn video, his hot aunt (by marriage, of course), his mom's best friend, the neighbor next door, the barista at the coffee shop, the college student at the bus stop, the girl seated across from him in biology class, and a scorching photo of Megan Fox — *all in the same day*!

Not much has changed since the classic 1969 novel *Portnoy's Complaint*, by Philip Roth, in which Alexander Portnoy, a sex-obsessed Jewish male, recounts his fantasies and exploits going back to his teen years, when he pleasured himself into his sister's bra, his baseball mitt, and a piece of raw liver. Thirty years later, in the film *American Pie*, high school senior Jim Levenstein (Jason Biggs) gets it on with warm apple pie — so at least teen males have moved on from the main course to dessert since Portnoy.

I admit that I was no exception to teen fixations many decades ago. When I was young, my good friend Keith lived on 21st Street in Santa Monica, California. As young men, we washed the Mercedes sedan and motorcycle belonging to the neighbor across the street, action film icon Arnold Schwarzenegger. One day Keith told me that Arnold had offered to lend us a porn flick and wanted to know what I thought. I enthusiastically said, "Oh yeah! We definitely need to borrow it!" That Saturday afternoon, Keith and I walked across the street to Arnold's house and rang the doorbell. Arnold opened the door, smiled, and said in his familiar accent, "C'mon in, guys. I'll get it." He returned with a brown bag, passed it to Keith, and said, "Enjoy, guys!" We walked back to Keith's house. When we entered his room, we opened the bag and pulled out a VHS tape of *The Opening of Misty Beethoven*. For us hormone-jacked teens, procuring this award-winning video was like winning the lottery. Later that night, Keith and I were in his family TV room while his parents and siblings were out. He popped the tape in the player and we were off to the races, so to speak. We enjoyed the film from beginning to end — after I paused and rewound it several times to figure out how the actors were able to contort themselves into such unusual positions — and eventually we offered our sincere thanks to Arnold when we returned the video.

I share the above story with you not to be vulgar but to underscore the point that our biological makeups coerce us at adolescence to seek out sexual content and satisfy our intense curiosities. Lest we erroneously think that teenage girls are immune to hormonal rage, consider the statistics from Covenant Eyes (an "Internet Accountability and Filtering Site") reporting that six out of ten girls have been exposed to porn before they're eighteen, and nearly a quarter of teen girls have seen bondage online. Teenage girls also masturbate frequently and have their own Portnoy-ish experimentations and fetishes; I've heard plenty of stories about women who have become intimate with certain objects found in the grocery store produce sections (zucchini, cucumbers, carrots…you get the picture).

MARK TWAIN RECEIVED QUITE A HAND FOR THIS SPEECH

In 1879 an internationally famous author and humorist delivered a speech in Paris entitled "Some Thoughts on the Science of Onanism" to an elite group of open-minded French intellects. The great Mark Twain — yes, the author of *Tom Sawyer*, *Huckleberry Finn*, and numerous other classics — spoke to his audience in satirical tones about the then taboo subject of, er, whacking off, with the clear implication that there is nothing wrong with doing what is biologically natural. The work, now available as a short illustrated book entitled *Mark Twain on Masturbation*, also presents his entertaining and modified quotes from historic figures. Twain joked that in the

second book of *The Iliad* Homer wrote: "Give me masturbation or give me death!" and that a Zulu hero named
·Cetewayo said, "A jerk in the hand is worth two in the
bush."

If we think these lines are funny and edgy today,
imagine what people thought of them in 1879! Certainly
the climate was more receptive to such talk in France
than it was in America at that time, and Twain was wise
enough to know how to safeguard his reputation and entertain only the right audience with such material. Even
so, Twain's ability to acknowledge and joke about masturbation during that conservative time period reveals PI
of the highest Jedi Knight order.

Does this mean that our Perceptual Intelligence level is dictated by how many films of sexual fantasies we have stored in our
brains? Deep down are we all perverts to some degree? I believe
this is only the case when we act on these fantasies and endanger
our secure lives with cloudy judgment and bad risk-taking. This
is no more true than in the case of infidelity — a temptation that
devastates couples with shocking frequency.

Illicit Behavior with a Hefty Price Tag

In the HBO TV series *Westworld* (based on the 1973 Yul Brynner
film written and directed by Michael Crichton, who had a fondness for action-packed theme parks gone awry), rich "guests"
have an opportunity to play out their deepest secret fantasies
in a Western setting with robot "hosts" who are so lifelike that
viewers have a difficult time figuring out who's who (or what's
what). With such freedom and abandon, male and female guests
can randomly have sex with (or rape) and torture (or massacre)
these robot hosts for whimsical gratification (and wads of untold

fees) — except that, well, at some point these hosts might remember the traumatic experiences after thousands of reboots, while simultaneously developing genuine emotions. The show is quite a reality/fantasy and ethical mind-bender to process, but here are the main questions for us to consider: *If a married man or woman has sex with one of the lifelike robot hosts, is it considered cheating? Are sexual thoughts of another person (nonrobot) okay? Is it fine to fantasize?* A man will never get divorced for thinking about a threesome with a supermodel, and few would argue that a wife who uses a dildo to achieve ten orgasms while fantasizing about George Clooney is cheating on her husband.

A staggering 41 percent of couples in the United States today admit that one or both partners has had an affair. Imagine how high that figure might be if all partners surveyed had been 100 percent honest. It's no wonder, then, that TV shows and films have tackled the subject of infidelity ad nauseam, and married audiences seem to love watching, despite the fact that the depicted affairs almost never end happily. Below are just a few representative programs and movies:

TV: *Scandal, Mistresses, Being Mary Jane, The Good Wife, The Affair, Divorce, How to Get Away with Murder*

Films: *Double Indemnity, Addicted, Fatal Attraction, Unfaithful, Tyler Perry's Temptation: Confessions of a Marriage Counselor*

In every adulterous situation above — unlike in *Westworld* — the guilty parties lie, cover up evidence, and combat severe bouts of guilt as they take part in their steamy, forbidden trysts. *Unfaithful*, which starred Diane Lane (Connie, the cheating wife) and Richard Gere (the cuckolded husband), raked in a whopping $119 million globally at the box office. This makes sense since the sex scenes are over-the-top raunchy, especially the unforgettable interplay in the hallway in which the lover (played by Olivier Martinez) enters Connie from behind as she tries to leave him (but can't because the sex is too good). It's not exactly a "princess" fantasy but one that titillates wives and husbands alike with its porn-like explicitness, danger, and exposure of getting it on in a public hallway. Despite the fact that things take a terribly macabre turn for the characters (I won't ruin it with a spoiler for those who haven't seen it), I would bet my bag of quinoa chips that most couples went home after the film, slammed and bolted their bedroom doors, and went at it like bunnies.

Do couples that indulge in hot sex with each other after watching a steamy flick like *Unfaithful* have low or high PI? I believe that it tilts far on the high side because couples capable of translating and incorporating fantasy into their romantic lives more often than not avoid acting on the unfaithful urges. The itch has been scratched, and both partners had a pleasurable peak in heart rate, with no harm done. Until a real Westworld Theme Park opens its gates — which may not be as far into the future as we think — let's agree to keep our relationships spicy and intact through activities that reflect high and healthy PI.

Not That There's Anything Wrong with It

Although this chapter has concerned Perceptual Intelligence in heterosexual relationships, since that is my own frame of reference, I would like to turn to the topic of the misconceptions and prejudice many straight people have against people with different sexual preferences.

When six-foot-four Hollywood leading man Rock Hudson died in 1985 from AIDS, the tabloids had an exploitive field day. As the first major celebrity publicly outed through this horrific (and, at the time, typically fatal) disease, Hudson's legacy fell prey to sordid headlines and overreactions from ignorant people who became shocked and repulsed that the manly image of the "Rock" — who had pursued actress Doris Day and other leading ladies on screen for decades — was forever tainted. In their minds he had shifted from Hollywood hunk to a symbol of all that was perverse in Hollywood (e.g., homosexuality). One can understand, given the extreme reactions, why Hudson had kept his private life deep in the closet all those years. As his professed true love, stockbroker

Lee Garlington said, "Nobody in their right mind came out.... It was career suicide."

Many other stars believed to have been gay, lesbian, or bisexual were able to keep their images intact by shielding themselves from exposure, including Montgomery Clift, Paul Lynde, Marlene Dietrich, Greta Garbo, Josephine Baker, Dick Sargent, Robert Reed, Raymond Burr, Katharine Hepburn, Randolph Scott, and Barbara Stanwyck. Homosexuality was much more common in old Hollywood than the public realized; the books *Full Service*, by Scotty Bowers, and *In or Out*, by Boze Hadleigh, provide striking personal anecdotes. Had tales of actors' sexual interests leaked out during their careers, their public personas would no doubt have been decimated, and fans would have turned against them.

Past (and sometimes present) generations, especially the ardently religious, had quite low PI, in that they saw homosexuality and lesbianism as threats to moral society and a danger to youth — as if sexual preference was some sort of contagious disease and could be dictated or controlled. More than a century ago, Oscar Wilde, the brilliant playwright (*The Importance of Being Earnest*), novelist (*The Picture of Dorian Gray*), poet, fairy-tale writer, and wit, was not only derided for his homosexual tendencies but imprisoned for them. His enclaves of fans and supporters abandoned him, and the artist died a broken man.

Thankfully, much has changed since the attitudes of Oscar Wilde's era and even in the years since Rock Hudson's tragic death. Due in no small part to the price their reputations (and others) paid, it's now common for actors, musicians, and artists to proudly come out of the closet without anyone batting an eyelash, and we have these and many other megatalents to thank for it: Ellen DeGeneres, Elton John, George Takei, Neil Patrick Harris, Wanda Sykes, Nathan Lane, Ian McKellen, David Hyde Pierce, Rachel Maddow, Simon Callow, Jodie Foster, Lady Gaga, and Lily Tomlin.

We've come a long way, but the world is not done with

misperceptions of and prejudice against the gay and lesbian community. In school yards you'll still see kids using words like *fag* and *lezzie* to taunt those they believe to be gay. There are few issues in politics more contentious than gay marriage, as if the formal ceremony and piece of paper will mangle our Perceptual Intelligence when it comes to the symbolic sanctity of marriage (whatever that means). In response to this ill-conceived train of thought, singer and novelist Kinky Friedman quipped: "I support gay marriage. I believe they have a right to be as miserable as the rest of us." There is no reason whatsoever to be uncomfortable with those of sexual persuasions different from our own, if we have high PIs and are comfortable with our own sexuality.

The Olympian Who Turned Heads — and PIs

In July 2015 *Vanity Fair* published a cover story on a woman named Caitlyn Jenner who had come forward as a transgender woman. We had all known her as none other than world-famous athlete Bruce Jenner: Olympic champion, product spokesperson, reality TV dad, and onetime *Playgirl* model. This unveiling was alarming to the millions of people who couldn't accept that Caitlyn, a woman, was the same person as the masculine and dynamic 1976 decathlon Olympic gold medalist captured in action on Wheaties boxes, which boasted photos of Bruce tossing a javelin. While family, friends, celebrities, athletes, and fans voiced and tweeted support, plenty of people were laughing behind Jenner's back — and a few in front of it. One TV anchor, Neil Cavuto of Fox News, went so far as to poke fun at Caitlyn on camera, subsequently referring to Charles Payne as "Charlene Payne." Carol Costello, a CNN anchor, wasn't much better, reflecting on her "sadness" that Mr. Jenner's manly physique would no longer be a sight to behold. Former Nickelodeon star Drake Bell made the fatal error of tweeting, "Sorry...still calling you Bruce," which caused an instant uproar in the online community.

Our PIs lead us to form impressions of people almost instantly, but they evolve and can get fixed over time. In Caitlyn's case, being a transgender woman is so far removed from the established male decathlon champion Bruce Jenner that some people to this day simply don't understand it, can't relate to it, and therefore will never accept it. Not that Caitlyn herself in any way requires their approval, and she has, in fact, been awarded various awards for her courage (which has also become a source of contention among some people).

The transgender issue, brought out in the open by Caitlyn Jenner, has stirred controversies about school bathrooms and public restrooms across the country. Should transgender men be allowed into the women's bathroom, and vice versa? If bathrooms are restricted to people who have the "right parts," who will be assigned the task of checking such things? It didn't matter to the sexually uptight people who were so concerned about this that they (and their children) had likely already used bathroom facilities frequented by transgender people yet never suspected a thing (and, most important, no incidents had ever occurred).

As a form of compromise, many schools and institutions have now created gender-neutral bathrooms with signs clearly demarking those invited to enter. Ironically, these are probably the cleanest and safest restrooms to be found. And to no small extent they have Caitlyn Jenner — a symbol of bravery to those who have PIs mature enough to accept her — to thank for it. Others who have low PIs will continue to hold on tight to a biased image: an Olympian who turned "superfreak."

Beasts, Anyone?

In Val Camonica, Italy, there is a cave painting dating from around 8000 BCE depicting a man about to penetrate an animal. In the 1950s sex researcher Dr. Alfred Kinsey investigated the prevalence of this sexual proclivity, estimating that 8 percent of men and

3.6 percent of women in the United States had sexual relations with animals, with a much higher percentage occurring among people who live on farms (especially acts with sheep). Statistics show that zoophilia (arousal from animal eroticism) and participating in bestiality (performing physical sex acts with animals) are more prevalent than ever, and words related to them are among the most searched porn terms on the Internet. This begs the question: Since it's been diagnosed as a biological desire and not pathological in most cases, do individuals who give in to such animal urges have low or high sexual Perceptual Intelligence?

At the end of the day, the mores of the respective social culture dictate what is acceptable and what is not. Bestiality is legal in much of South America and also in countries as diverse as Russia, Mexico, Thailand, Japan, and Finland. Denmark recently passed laws against sex with animals. The United States is all over the place with the issue; some states deem it a misdemeanor, whereas others, such as Texas and Kentucky, treat it as legal. It's not a mere coincidence that a lot of "zoo porn" originates in Russia. In these videos, Russian women get it on with dogs, donkeys, horses, and assorted other animals. Russian culture has been relatively accepting of such behavior over the centuries (though there is no truth to the myth that Catherine the Great died having intercourse with a horse), so it's become embedded in their psyches and is thus more out in the open and available for viewing.

One major consideration is whether or not bestiality is considered animal abuse. The case is pretty straightforward when an animal exhibits pain or suffers an injury from an encounter. But what if no such sign is exhibited; is it always considered a violation because the animal isn't known to be a "consenting" partner? Then again, isn't it the case that among some animal species the males enter females without requiring permission?

Until Dr. Doolittle himself shows up to translate barks, bleats, brays, and neighs, we are left in the dark on the matter. One thing is for sure: those who have this sexual proclivity and have uncontrollable urges to act on it are best served ensuring that no creature gets harmed and that the barn doors are tightly secured to protect everyone's privacy and virtue, including those of the animal. No one wants to see a "scarlet A" on Bessie the cow.

Alas, we must depart from this chapter of iniquity and shift to why we desperately long for things that are expensive, rare, and beyond our reach. One of these is not a prized sheep but something perceived as invaluable that happens to originate from the back end of a very different animal.

11 Gotta Have It

Cat Poop Coffee and the Seduction of Low Production

Have you ever desperately longed for something (or someone) but were aware that it (or he/she) might be just beyond your reach? It could be an expensive new car, a piece of jewelry, or an attractive person who plays hard to get. Obtaining the object of elusive desires can be accomplished with great effort, but why are you going there in the first place? Does the prize have the same merit as your perceived value of it? Will having it demonstrably improve your quality of life? Or have you simply been swept away in the fog of lusting after something you can't have?

It's a basic tenet of human behavior that we desire a thing more when we perceive it as rare or difficult to obtain. An opportunity may seem more valuable if we are convinced that its availability is limited. Indeed, study after study has shown that when we view something as rare, unique, or hard to get, our perception of its value and attractiveness skyrockets. Even those with high degrees of Perceptual Intelligence may become victims in this arena.

At a London coffeehouse, people pay the equivalent of $100 for a cup of Kopi Luwak coffee. This sounds exotic, delicious, and like a terrific pick-me-up to go with an oat bran muffin in the morning — but what is it? Brace yourself: it's *cat poop coffee*. Yes, you read that correctly. Kopi Luwak coffee is produced in the following unusual (or gross, depending on your point of view) manner. An odd nocturnal Asian animal named the Asian palm civet (a.k.a. "the toddy cat") eats whole and digests the fleshy pulp of coffee cherries. The product coming from the tail end of these rare, cute, and sometimes mistreated creatures is picked off the ground. The fecal matter is cleaned off (I wouldn't want to have *that* job), and the beloved coffee beans are sold at a premium around the world. Before you go sifting through your cat's litter box for some steamy chunks, be warned that the Asian palm civet has relatively nothing in common with domesticated cats. To me, the animals more closely resemble hybrids between lemurs and raccoons.

In chapter 7 we pondered the perplexing question of how and why someone would pay $28K for a grilled cheese sandwich. Is there a PI connection between this grand misperception and when people pay through the nose for aromatic cat poop coffee? The two are related, but different thoughts, emotions, and delusions are equally disturbing with relation to the latter extravagant, completely unnecessary purchase.

I Want It and I Want It *Now:* Cabbage Patch Dolls and Beanie Babies

The "rarity" of an item artificially increases our desire to have it while adversely altering our judgment. When that prized image hits us, we revert back to childlike wants and desires, and all sense of perspective goes out the window. Scarcity works when people perceive that something is more valuable if it is expensive and/ or less attainable. Usually those with low PI are likely to succumb and make the purchase because they fail to have insight into what is at work in their minds. However, even individuals with high PI on most matters struggle when that overpriced, bright, and shiny desirous object strikes their fancy.

Obvious examples of the above are items like gold and silver, which can be crafted into an unlimited number of beautiful shiny objects. These metals have been sought out and cherished since the beginning of human creation because they don't magically fall from the sky. If there were aliens on another planet that could grow and cultivate gold and silver like we do sugar cane, I'd wager that these precious items would be considered worthless, or at least as valuable as gummy bears.

Often people hear or read about a certain tantalizing product without having all the information about it, such as whether it's well manufactured or worth the retail value. Therefore, they rely on shortcut cues — such as how expensive or limited it is — to determine its value. If everyone is clamoring for a trendy new Christmas toy, then it must be good (even if in reality it's mediocre, or worse). Many will remember the intense fervors caused years ago by the crazes for Cabbage Patch dolls and Beanie Babies. During the height of these fads, what made the items seem so essential to children — as well as to their parents, who considered them collectibles — was their limited availability, their soaring prices, and the perception that those who had them possessed something of high worth. If a youngster didn't receive the right Cabbage Patch doll or Beanie Baby as a gift, there would be tear storms, not snowstorms, for the duration of the holiday season.

The intensity adults feel about owning an object, whether it's a new sports car, an iPhone, an elaborate flat screen TV and sound system, an expensive necklace, or a cup of cat poop coffee, is no different from what goes on in the mind of the child desperate for that Cabbage Patch doll. Once it's set in our brains that *we've gotta have it* because the object is rare, costly, status driven, or fought over in stores by desperate customers, our desire becomes all consuming and our low PI fails to protect us — or our credit card statements.

The brass ring becomes even more valuable if it's forbidden and/or somehow dangerous. Teenagers have pursued relationships and things they were told they couldn't or shouldn't have *because* they were restricted. Romeo and Juliet became infatuated with each other at least in part because their families were involved in a blood feud. They were well aware that their relationship was dangerous, and this added to the excitement of Romeo wooing his dream girl at her balcony. Had the Montagues and Capulets vacationed together at a time-share in Naples, I doubt their kids would have been as drawn to each other. Telling a teenager not to smoke pot often has the unintended effect of increasing the intense curiosity and desire for a joint. (This doesn't mean that I advocate your doing bong loads with your teenagers.) If a book is banned, we're more likely to go out and buy a print copy (or download it, for greater privacy).* A jury member instructed to disregard evidence will probably cling to the idea of getting that information anyway; her wondering what it might be could soil her view of the entire case. When we are caught off guard with low PI, we often cave in to our curiosity.

Why Doesn't Junior Play with His Toy Anymore?

You've finished swilling down your $100 cup of cat poop java, that is, Kopi Luwak. You've driven your hot new Ferrari for several screeching spins around the block. You've shown off that limited-edition Tiffany diamond necklace at your best friend's wedding. You even waited in line overnight to get the new iPhone the day it was released.

If you celebrate Christmas, this is the typical postholiday scenario: the area around the Christmas tree has become a battlefield of torn wrapping paper, boxes, plastic covering, and unopened instruction manuals. Somewhere strewn about are the lonely and forgotten toys that only a little while earlier had been unwrapped

---

* In the 1960s and 1970s, when films like *I Am Curious Yellow, Caligula,* and *Deep Throat* were given X-ratings and banned from theaters, you can bet that people put them on their lists of must-see movies. (A humorous twist occurred when some clever marketing guys working for Plymouth named one of their 1970s muscle car colors "Curious Yellow," which is believed to have been a joke on corporate management, since it flew under their radar and got approved.)

to exclamations of joy: "Mom, you're the best! This is the happiest day of my life!"

Whether you were that child with a toy or an adult who has longed for an out-of-reach item, the aftereffect is universal. It may take a few months, hours, days, or even minutes, but once the reality of ownership sets in, we may start to feel a letdown. Children tend to get easily bored with toys, but adults who have heightened Perceptual Intelligence return as if awakened from a dream and, for the first time, objectively see the true value of their purchase. The stronger the individual's PI, the sharper the buyer's remorse afterward. *Wait,* you think to yourself in disbelief. *Did I just shell out a hundred bucks on a cup of cat poop coffee I'm only going to pee out? I'm such an idiot. That was totally not worth it. And, oh man, my stomach is gurgling from it already. I wonder if I'm going to have an urge to pee in the dirt and cover it over with my paw!*

The sad truth is that our PI and emotions were fogged up at the time of the purchase and only now self-corrects. Once the prize is in our possession, we come to our senses (but not always) and realize what we've done. Invariably, we feel stupid, wasteful, embarrassed, disappointed, and rooked. We inflated the value of the purchase to such a ridiculous proportion that the reality can never match the grandiose illusion.

Hopefully we'll make a better decision the next time, right? Wrong! The feeling at the time of purchase was so exhilarating, owing to the dopamine blitzkrieg in our brain, that we have forgotten how foolish we were on that last occasion. We subconsciously only remember the wonderful childlike feelings of *wanting* and *having,* and we ignore the dent the earlier expense made in our wallets and all the disappointment we felt afterward. The lesson isn't learned because we enjoyed those feelings so much that we again receive childlike signals and images we can't resist. Our PI is so low that we rationalize: *The cat poop coffee? This new thing is nothing like that, it's totally different. It's even better! This time I'm getting cat poop espresso — I just have to try it!* That's called rationalization of buyer's remorse.

The best way to avoid such temptation is to leave your credit cards with your significant other or a close friend for safeguarding when heading into your favorite store. You can always take a picture of the item and purchase it later when your PI has simmered down. Time is your friend: the more time you are able to spend away from the product, the greater chance you have to avoid impulse. You may also wish to consider spending less time trolling websites, where purchases are just a few clicks away.

Illusions of Godlike Power

We all enjoy a gag gift or novelty item from time to time, but how some of these bizarre items catch on and become multimillion-dollar fads sometimes defies logic. Our fascination with these strange objects is deeply rooted in our childhoods. Many of us growing up in the 1970s have fond memories of flipping through the pages of *Archie*, *Richie Rich*, or *Spider-Man* comics and getting immersed in ads for some pretty outlandish things. Right there, for only a buck and a quarter, you could purchase an entire aquarium filled with *sea monkeys*. These remarkable "instant pets" — depicted with grinning alien-like faces and human bodies in the print ads — would seemingly come to life out of virtual nothingness. All you had to do was add tap water and voilà! — these mysterious beings would be swimming around in a tank right from birth, ready to be "trained."

Imagine: if you're a child whose parents have refused to get you a dog, a cat, a bird, or even a gerbil, you could fork over a paltry $1.25 to breathe life into an entire sea monkey universe. Few children at the time realized that the sea monkeys were nothing more than brine shrimp mixed with chemicals to form a sustaining environment until the water was added to "reanimate" them. I'm sorry to burst the bubble of anyone who still believes in sea monkeys, but trust me when I tell you it was all just a hokey science experiment designed to snag your hard-earned paper route money. Last I checked, these creatures are as trainable as plankton — and just about as cuddly, too.

When we are children and our Perceptual Intelligence is low, we are easily susceptible to such hype. Our imaginations and unformed self-perceptions are already running wild, so we are fooled into accepting that it is entirely possible for us to have the power to *create life-forms* and then to claim them as *pets*.

Delusions of grandeur were not the only reasons these goofy comic book items seduced us. Some people may fondly recall the infamous ads for special "X-Ray Specs." For only a buck, a whole quarter less than the menagerie of sea monkeys, you could allegedly see through people's clothing and catch peeks at their naked flesh — quite a promise, especially to hormone-driven teens back in the days before the widespread availability of Internet nudity. The product, from Harold von Braunhut, the wealthy "creator" of sea monkeys and other gimmicks, was a complete hoax (the lenses were *made out of cardboard* with tiny holes to see through) that roped many youthful buyers into thinking that not only was it possible to own glasses that bestow upon the wearer X-ray vision but that such incredible power could be had simply by spending an amount that happened to be the rough equivalent of a weekly allowance. Those suckered into buying the X-Ray Specs may have been gullible, but they were not stupid. They had low PI from having been in this vulnerable state of mind (I'm sure some adults ordered them, too), and were fooled into thinking X-Ray Specs could help them see through clothing. The answer to the question of why this was so is simple: the longing for the perceived glimpses underneath people's garments was so intense that logical thought abandoned them, replaced by fantasy.

On a deeper level, the glasses would grant them the fantasy of having superpowers and perhaps even give them other hidden abilities not mentioned in the advertising. The power of testosterone driving X-Ray Specs sales cannot be underestimated. Consider the sheer tsunami of online porn cited in chapter 10.

You Gotta Have a Gimmick: The Big Mouth Billy Bass Fish

Alas, we're still vulnerable to spending our money on novelty items and cheaply produced goods even as we get older and allegedly wiser. We become targets for all kinds of manipulation and exploitation as we continue to fall for inane tchotchkes throughout our lives, whether or not we are aware of it. We may be incredibly smart and savvy in any number of areas yet still have low PI when it comes to distinguishing *wants* from *needs* and keeping our impulses in check. In some cases, buying stuff fills a void in our psyches left over from a deprived (or spoiled) childhood. In other cases, as I'll explain below, the hype surrounding these items validates them in a way that appeals to the childlike aspects of our nature.

Some of these alluring objects, such as false teeth and the classic Groucho glasses, have become timeless gag gifts. More recently, cheesy noisemaking gag gifts like yodeling pickles and the infamous Big Mouth Billy Bass singing a rendition of "Take Me to the River" have shown up at holiday parties and birthdays, inducing gales of laughter when unwrapped.

Once in a while buying gag gifts to give as a joke or to add some fun to a get-together can be pretty entertaining. Yet how on earth could anyone be coerced into purchasing dehydrated water ("just add water"!), a chia pet, a leg lamp (shades of the film *A Christmas Story*), a pet rock, a toilet golf game, or an umbrella hat? Believe it or not, entrepreneurs have done quite well from these gimmicks, and they must have had at least some inkling that customers existed who would be willing to purchase them.

Advertising for novelty items buries your skepticism and lowers your PI. Chances are you saw the item advertised many times on an infomercial, as an online pop-up, in Spencer's gifts, or on TV. The pitch lines "over five million sold" and "limited offer" trick your brain into taking impulsive actions. You *have to have it* because it's been validated by all the repeated exposure and endorsement by the number of other buyers. You have an irresistible urge to show it off to your relatives, buddies, and neighbors. *In fact, you think, doesn't Uncle Bob fish and wear goofy fishing hats? He'd crack up if I got him that Big Mouth Billy Bass singing fish for his birthday!*

We've all been in the position of showing off a silly purchase — either for ourselves or as a gift — and being met with eye rolls and chuckles. Our friends don't get it at all. For those with a modicum of Perceptual Intelligence, this smacks them into reality and causes their cheeks to turn beet-red with embarrassment. Those with low PI probably don't notice how others react to their basement, garage, or office, overloaded with these "collectibles," which might fetch pennies at their next yard sale.

The latest craze among kids as of this writing is the fidget spinner, a plastic hand toy designed to alleviate boredom that is also purported to help reduce stress and fidgeting while improving focus. If the device is being purchased because "everyone has it," this indicates low PI; if it's to assist with the aforementioned issues and shows some genuine benefit, then the PI equation shifts in the other direction.

The Low PI of Being a Tourist

The must-have scenario is especially prevalent when we go on vacation. As we set off for much-needed rest and relaxation, a switch gets flipped and we soften up so much that we become susceptible to spending money on all kinds of things we normally wouldn't buy. Most of us have really low PI when it comes to overindulging our "vacation brains."

This scenario is probably eerily familiar. The family has driven just an hour and a half on their way from New York to Florida and already needs to take a break at a rest stop because junior has to pee *really bad*. The family flops out of the SUV and heads into the public facilities. On the way out, each individual meanders through the gift shop — which is always right next to the lavatories — waiting for everyone to be ready. Before you know it, Mom and Dad are at the register buying stuff for each family member: a magazine, a crossword puzzle book, Mad Libs, a magic trick card set, a whoopee cushion, a snow globe, a chia pet...*whoa*, what just happened?

The instant we set off on vacation we are swept away by the relaxing vibes. The stress of work slides off our shoulders and we let our guards down. We want everyone to be happy, and spending money on unnecessary things may fill that need. Plus, the kids are kept busy (we hope) with all the stuff and won't bother Mom and Dad or fight with each other when the Disney video ends.

This vacation-induced low PI in all likelihood won't improve as our trip progresses. In fact, our PIs often get *reduced* the longer we are away. We might resist the fancy watch in the window the first time we see it as we strut around town in our shorts and T-shirts. But the image of the watch never quite leaves our minds, and we start to think that we might not have another chance to buy it — especially at that "sale" price. By the second or third time we stroll past that window, the watch starts to "call to us," and we rationalize: *What the hell — it's vacation!* We're lost in the "vacation zone."

The shopping compulsion that overtakes us on vacation isn't limited to bribing the kids to behave with shiny new doodads. If we aren't in tune with our Perceptual Intelligence, it can plummet and make all of us receptive to the warm, fuzzy feelings of vacation that lead us to buy items connected with our time away. People on vacation love to shop and buy. When we aren't snorkeling or sunbathing, we frolic through all the local novelty, specialty, and clothing stores without a care in the world. At first it starts out as a fun thing to do, but what we don't realize is that our subdued PI has failed us as we drift into a calm dreamland where spending is of less concern and causes us to let loose. No one likes the idea of going off on vacation and coming back empty-handed, including things for oneself, not just for those back at home. Why not? Everything in the local gift shops looks so much better than the stuff back home — or does it? And so you buy (or "borrow") bathrobes from the hotel and purchase themed mugs, shot glasses, shirts, and so on. Anything that smacks of being authentically "local" — the maple syrup in Maine, the rum in Bermuda, the hot sauce in Louisiana — takes on the illusion of being unattainable back home or better when bought locally and therefore becomes highly desirable. How good does your beach vacation purchase of the multilayered-colored-sand-in-the-perfume-bottle really look in your bathroom?

We also become suckers to the kids who "need" souvenirs and mementos at every pit stop: in New York City, it's a mini Statue of Liberty; in Philadelphia, it's a tiny Liberty Bell; in Boston, it's a Paul Revere coloring book; in Washington, DC, it's a puzzle of the presidents. (Yes, I admit it: I'm guilty of these PI felonies with my daughters.) We can all rationalize that these items have some potential educational value but, at the end of the day, it's the experience of being *somewhere else* that fogs our perceptions and tells us it's all right to plunk down our hard-earned money on stuff.

I don't insist that you lock up your credit cards in the hotel safe when you are on vacation (although for some people, such as those touring Las Vegas, that's not a bad idea). It's perfectly fine to treat yourself every once in a while and to cave in to the illusions of "needing" something, as long as you're spending within your means. (As a suggestion for your next vacation: if someone amped up on happy pills invites you to attend a time-share presentation, I suggest you attempt to beat Usain Bolt's hundred-meter world record while you run in the opposite direction.*) It is important, however, that when you're in that colorful gift shop, you think through whether or not you're being lured to buy because of the surroundings while your guard has been lowered. Ask yourself this: If that same object were available on a shelf at your local

---

* You may thank me now for this advice.

gas station mini-mart, would you still plunk down the money to purchase it?

In the next chapter we'll discover how groups of people can influence not only your spending habits but also your decision making and behavior across the board.

12

Are You Different from a Wildebeest in Kenya?

The Dynamics of Social Influence

When I was in Kenya I had the opportunity to witness firsthand a unique form of peer pressure. It was while I was on safari observing a kind of bearded antelope called wildebeests, large animals known for having remarkable appetites. I studied a portion of some three and half million of them as they moved as a herd from the Masai Mara savannah in Kenya to Serengeti, Tanzania. Suffice it to say, lions, tigers, cheetahs, and other predators anywhere in the vicinity must have been licking their chops at the sight of the hearty buffet marching right out in the open. Even so, perhaps there was one freethinking wildebeest among the group who declined to join the others and said, "Uh, no thanks, guys. I'd rather not become someone's wildebeest loaf."

Whether or not you realize it, we have a lot in common with the majority of these wildebeests who blindly follow the herd. (I'll skip the DNA discussion for another time.) Although you walk on two legs and have thumbs, don't think you're so superior to those four-legged beasts. We, just like the wildebeests, are affected by social influence from the time we are born.

When you were a toddler, you laughed because you were following suit when your mom, dad, and older sister laughed aloud, even though you didn't get the knock-knock joke. When you were in grade school, either you joined in on the swirl of taunts against a classmate who tripped and fell in front of everybody — spilling a drink on the most popular girl in class — or *you* were that unfortunate kid and others hurled those jibes at you. When you were older and at a college party, you hesitated at first but ultimately joined your pals in a campus crusade against an exposed unfairness, although you didn't quite understand what the cause was about. As an employed adult, you sat in a team meeting with your boss and nodded your head in agreement with everyone else to the boss's poorly conceived proposal, though deep down you had some reservations about her course of action.

These follow-the-leader responses are triggered by many emotional needs, the main ones being fear, risk aversion, and the desire for acceptance. Conforming brings us immediate personal validation; we feel good about being part of a collective or perhaps relieved that we won't get our heads handed to us on a plate by our peers or supervisor. Psychologically, when we concede to the majority, we start to rationalize: "Well, so many people agree with it…maybe they're right and I'm wrong to think otherwise." Sometimes it just feels safer to go with the flow.

Having low PI with regard to social influence can be a blessing or a curse, depending on the circumstance. If you have low PI and are the college student who joins the bandwagon to petition the school to correct gender imbalance among the faculty (even though you've never really given it a thought), then you have lucked into being on the right side of history. If, however, you have low PI and go with the flow to join the battle of organic cheese versus the GMO cheese served in the school cafeteria — which leads to hurling grilled cheese sandwiches at the dean's office building — well, then, your low PI has led you down the path of Bluto from *Animal House*.

In his book *Influence*, Robert Cialdini asserts: "Without question, when people are uncertain, they are more likely to use others' actions to decide how they themselves should act." Cialdini humorously refers to this as "monkey me, monkey do." Smart marketers and advertising people have field days playing on our low PI in this regard. Having a high PI buffers people against this, since they have self-awareness and insight into this "monkey see, monkey do" phenomenon. On the other hand, shoppers with low PI are apt to select a product or service with a larger number of followers and endorsers. If a TV commercial depicts fifty thousand people in a sports stadium guzzling Pepsi and then cheering about it, people with low PIs watching from home start to subconsciously think, "Coke may not be *it*" any longer and become receptive to trying out the competitive product.

In a 2014 study, "Social Defaults: Observed Choices Become Choice Defaults," researchers refer to an automatic process known as *behavioral mimicry*. Picture two Chinese restaurants side-by-side. One is packed and the other is empty. With all things being relatively equal, such as menu offerings and pricing, which one will you decide to enter? Odds are you will head into the more crowded establishment only because it has been validated by the presence of more people. Virtually all of us do this. You probably won't consider the possibility that the crowd could have been driven by a Groupon promotion or some other hidden factor. Why does it rarely cross anyone's mind that it might be more beneficial to go to the less crowded restaurant because the service would be faster and better, given that they have fewer customers?

In all likelihood, no one gave you the advice that it's okay not to follow the herd. If anything, you've been pressured to conform, and to feel anxiety at not conforming, by a large group even when nothing has been stated outright. With that being the case, how do you know when it actually *is* best to conform? In the examples that follow, we will see that having low PI can become either a positive or a negative factor in our judgment, depending on what kind of social influence is being engaged.

We Are Stardust, or We Are Pure Lust:
The Children of God (1969) vs. Rage Against the Latrine (1999)

If you are of the baby boomer generation, chances are at some point in your life you listened to the 1969 Woodstock Festival soundtrack, watched the Academy Award–winning documentary film, or both. Or maybe you or someone you know attended the event sprawled across farmland in Bethel, New York, and witnessed musical history: thirty-two of many of our greatest modern performers playing in front of some four hundred thousand people (when only about forty thousand were expected). Aside from the timeless music generated by legends such as Jimi Hendrix, Janis Joplin, Richie Havens, the Who, Joe Cocker, Sly and the Family Stone, Arlo Guthrie, and Joan Baez, by all accounts the event should have been an unprecedented disaster. In addition to the pervasive drug usage and sloshy mud from ill-timed rain, the promoters were grotesquely unprepared for the scores of people, who gummed up highway traffic, poured into the surrounding areas, and trampled across the farmland. There was a shortage of medical care, food, water, and bathroom facilities, while such essentials as drugs and libido were in high supply. Stagehands were scared to death that a performer might get electrocuted from all the rain drenching the speaker wires. Miraculously, the damage incurred during the festival was minimal: two deaths (one a heroin overdose, the other a teen in a sleeping bag who was accidentally run over by a tractor) and eight miscarriages.* In spite of all the things that could have gone terribly awry, the festival indeed lived up to its reputation as "three days of peace, love, and music." The only incident was a tirade on stage by activist Abbie Hoffman to free jailed pot smoker John Sinclair, which resulted in his getting clobbered on the head with a guitar by Pete Townshend,

---

* There may have also been one to three births, but no one has officially stepped forward as a "Woodstock baby."

the Who's famed guitarist (who more typically seemed to enjoy smashing his guitars on stage while playing them).

Flash-forward thirty years to August 1999, Rome, New York, where the third incarnation of the Woodstock Festival was held and featured such superstars as Counting Crows, Sheryl Crow, Dave Matthews Band, Red Hot Chili Peppers, Kid Rock, Metallica, and Alanis Morissette. While the attendance was half that of the original Woodstock (a paltry two hundred thousand) and the organizers had the benefit of hindsight from the two prior Woodstock events, not to mention many other large-scale festival concerts, this one became an all-out debacle. Although no one can be held accountable for the blistering hundred-degree heat, someone probably should have sensed that the summer temperature would be a problem and provided more protective shelter areas against the sweltering sun. The organizers might also have realized that charging $4 for a bottle of water when there were few other liquid alternatives was a pretty inhumane idea, despite the allure of having a captive audience. Seven hundred people ended up being treated for dehydration, since the water fountain lines were miles long; many of these fountains were destroyed out of frustration. But that was only a portion of what hit rock bottom: riots and looting broke out, areas were torched, thirty-nine people were arrested, at least four women were raped (including a reported public gang rape), and one individual died of symptoms related to heat stroke.

Why was the first Woodstock triumphant against its hardships but the third such an all-out failure (except for the music)? The aura of the first Woodstock in 1969 — which translated from the creators to the promoters to the workers to the musicians to the Bethel police, volunteers, townspeople, and many other unnamed heroes — was uniformly about the *peace, love, and music* mantra. Yes, one could say that the naïveté of the era, the explosion of innovative rock and folk sounds, and all the mind-distorting drugs played major roles, but I also believe that something else prevailed that did not in 1999: *positive social influence.*

The hippie culture of the late 1960s promulgated pacifism, but preaching it and doing it are two entirely different things, especially when conditions go south. So why were there no riots, rapes, arson attacks, or arrests on that farm in the summer of 1969? The communal feeling had already been established before the event, starting with the creators, and everyone climbed on board with passion and fervor. The message to stay chill and help out their fellow brothers and sisters was contagious. Not only was the first Woodstock free to concertgoers to prevent tensions from mounting (thus ending up a financial bloodbath for the investors), but the entire town and neighboring areas pitched in without much complaint to get through the three days because the hippies were peaceful, were mellow, and didn't cause any trouble (except for leaving a lot of garbage behind). On all sides, the social influence of peace, love, and music had consumed them because it had been shared between one individual or group and the next. Those involved were not aware of it, but the collective low PI of peaceful conformity helped the situation, despite all the negative circumstances that could have turned the event into a modern Pompeii.

SLY STONE'S HIGH PI

One highlight performance of the 1969 Woodstock Festival was by Sly and the Family Stone. Bandleader, singer, composer, and organist Sly Stone gave a powerful introduction to the song "Higher" that displayed his gift for generating audience participation. He even began by astutely acknowledging that some people are uncomfortable with sing-alongs and need permission to sing in public, and so he assured those with high PI that it was okay to join in. I can't pretend to capture all the style and funk, but hopefully you'll get the gist of it:

We're gonna try to do a sing-along. Now, a lot of people don't like to do it because they think it might be old-fashioned. But you must dig that it is not a fashion in the first place. It is a feeling. And if it was good in the past it's still good. We're going to sing a song called "Higher," and if we could get everyone to join in, we'd really appreciate it.... What I'd like you to do is sing "Higher" and throw the peace sign up — it'll do you no harm. Still again, some people feel that they shouldn't because there are situations where you need approval to get in on something that could do you some good.... If you throw the peace sign up and sing "Higher," get everybody to do it.

If you are so inclined, go on YouTube and check out the charismatic singer and his "family" in action. You'll discover how an entertainer can create unity and a historic moment by using social influence techniques to appeal to a collective Perceptual Intelligence. Now *those* are good vibrations!

By contrast, the 1999 event was moneymaking commercialism at its worst. In addition to the $4 charged per water bottle, the event organizers seem to have given little or no thought to the scalding-hot tarmac or to providing enough basic services to the attendees. Even knowing the lessons of past Woodstock festivals, there *still* weren't enough toilets. From the top on down, no one seemed to care about what went on, and barbarism along the lines of William Golding's novel *Lord of the Flies* set in. The unruly behaviors spread like wildfire among those with rock-bottom PI, including the performers. According to the *San Francisco Examiner*, Kid Rock spat out a lame attempt at making a statement about recycling as he "demanded that the kids pelt the stage with water bottles." It had more than the desired effect and riled up the parched, pissed-off crowds, who followed his suggestion. But the audience had its own swirling negative influence going on, such as when they inappropriately launched into singing "The Star-Spangled Banner" during the Canadian band Tragically Hip's rendition of "O Canada." Unlike the first Woodstock — which warmly accepted Ravi Shankar's music alongside the 1950s-style vocal group Sha Na Na, the 1999 festival would be remembered in infamy for its greed, lack of humanity, offensiveness, and violence — not for the music.

We've established that charismatic performers can influence significant communities of people who have either low or high PI that achieves either positive or negative ends under different conditions. I don't need to get into specifics to point out that dictators and tyrants from Julius Caesar to Attila the Hun to Napoleon to Hitler to Putin appeal directly to followers at the bottom of the Perceptual Intelligence scale, typically rallying naive swarms with a message of unification, resulting in mass destruction. But what of the hucksters, tricksters, and con artists who fool hundreds, thousands, and maybe millions all at once in less directly violent ways? Where do they fit into the PI landscape?

A Million Suckers Are Born Every Nanosecond

What do Steven Spielberg, Larry King, Sandy Koufax, Elie Wiesel, and John Malkovich have in common? These are just a few high-profile individuals who were (or still are) blessed with intelligence, fortune, fame, and talent. These figures were considered to be at the forefront of their fields while in their prime, and some might say they are among the most gifted of all time. And yet all of them — as well as hundreds of other smart people and well-protected major organizations — were made to be fools, ripped off by one notorious individual in whom they had placed their wholehearted trust: Bernie Madoff.

Before the turn of the millennium, no Wall Street financier was more revered by people in the know than Bernie Madoff. The investment kingpin who created a fifty-billion-dollar Ponzi scheme over a period of many years had pulled the wool over everyone's eyes by creating an image of himself as a wise, kind, gentle, and generous man who made friends instantly, was well liked by his employees, and seemingly gave with all his heart to charity and community. With a multidecade performance worthy of a Best Actor Oscar, Madoff was considered beyond reproach in and out of his circles, including by elite country clubs, religious institutions, powerful corporations, international banks, and everyday investors. Madoff could have been Mother Teresa's second brother.

Psychologists and social scientists believe, in retrospect, that Madoff was pathological. He may even have fooled himself into believing that he had been "doing good" because he was universally liked and he donated to charity. Madoff likely had low PI when it came to distinguishing reality from fantasy, but without a doubt he used his charm to appeal to the even lower PI of those who blindly fell for the facade of *social proof* (as Cialdini referred to it) when investors didn't know what to do with their money (a.k.a. *uncertainty*, to borrow another word from Cialdini). To take this a step further, I believe that this low PI coalesced into a massive shared hallucination. The well-known folks cited above are not financial wizards who could have analyzed Madoff's activities to determine competence. They naturally took the shortcut by relying on how their peers were investing their money. When the curtain was pulled back and Madoff was exposed as a charlatan, the world awoke as if from a shared nightmare — their fortunes devastated — and collectively thought: *How could we have been so stupid? I thought he was my friend!* Even famous people can have low PI in areas outside their professions.

P. T. Barnum famously said (or famously didn't say, according to some sources), "There's a sucker born every minute." I think it would be more accurate to say that a *million suckers are born* *every nanosecond* because we admire and long to follow someone who is successful and self-made, seemingly guides others do the same, and exudes a saintly persona. Most of us have low PI to some degree in this regard, longing for that one person to exist whom we can trust. One monkey thus follows the last monkey, who follows the last, and so on until there is an endless amalgam of intertwined monkeys — helpless victims waiting to be blasted inside the barrel.*

Steering the Herd: Influencing the Masses with Celebrities and False Advertising

In the classic Alfred Hitchcock film *North by Northwest*, Cary Grant plays an ad man who defends his occupation by saying it's not lying but "the expedient exaggeration of the truth." We accept this as part of our cultural fabric and, in many cases, are entertained by representations of slick advertising people, such as in the phenomenally successful TV show *Mad Men*. But when does this "expedient exaggeration" end up crossing the line?

Consider the following examples from years past:

- Rice Krispies "has 25% Daily Value of Antioxidants and Nutrients."
- ExtenZe has been "scientifically proven to increase the size of a certain part of a male body."
- Rogaine is the number one dermatologist-recommended brand for hair regrowth: "The active ingredient in Rogaine products, minoxidil, reinvigorates shrunken hair follicles promoting hair regrowth and thicker hair over time."
- Splenda is "made from sugar."
- Activia is "clinically" and "scientifically" proven to be healthier than other yogurt.

---

* Barrel of Monkeys is a popular game you might recall in which you linked the monkeys' arms together. The plastic primates came in a distinctive barrel.

- Kashi's cereal products are "all natural" and "contain nothing artificial."

Millions of people bought the aforementioned well-known products because of this advertising, but *only one of these slogans is believed to be true*. All the products except one were slammed with false advertising claims and lost tons of money in legal fees and fines. Can you guess which one it is? All right, I'll alleviate the suspense: Rogaine was the only product from the above list to meet its advertising claim. Did you name it correctly?

I do not mean to suggest that Rice Krispies, Splenda, Kashi, and so on are good or bad products but simply that their advertising in these cases was misleading. The advertisers used trickery and exaggeration to fool large groups of people, and this is problematic for those with low PI — which could be any one of us when we "need" or want something, especially when we are rushed or stressed by our daily lives.

We are faced with "expedient exaggerations of the truth" and false advertising everywhere we look. It's particularly difficult for groups like the Food and Drug Administration to monitor and keep up with such activity on the Internet, since items can pop up and disappear in the blink of an eye. With that in mind, consider these questions: What was the last thing you purchased that didn't live up to its advertising? What lured you in to buy it? Did you return the product and go so far as to write a letter to the company, or did you let the matter go?

The main way to protect your PI (and your health) is for you to keep your eyes wide open whenever you are shopping. Be skeptical. Don't immediately believe throwaway bursts on packaging like "all natural." Always read product labels carefully and know what you are really buying. Products may promise things like "no MSG added" — but does that mean there might still be some MSG already embedded in the food and that's why it didn't need to be added?

So far in this book I've focused primarily on the negative

aspects of the Internet, but there is good, immediate information available there as well — as long as you are certain the sources are reputable. Many sites feature user reviews on the products show-cased, which can also be helpful in making purchasing decisions. I advise a bit of caution here as well and suggest not taking every review seriously: some may have hidden biases one way or the other (they may be written by friends or family of the manufac-turer or, on the other side of the spectrum, by a competitor), and there have been instances of reviewers who wrote their commen-tary without even having used the product or service. Also, some people just like to write crazy stuff on public sites just for the hell of it.

Trust in Me: Why Certain Actors Are Chosen for Advertising

It's no secret that advertisers are professionals paid to manipulate the public into buying products and services. Many companies use celebrities to tout their wares. The Celebrity DBI from the Marketing Arm ranks celebrities based on trustworthiness. Tom Hanks has ranked high on the list for years. If he appeared in a commercial purporting the benefits of a horse hair–based supple-ment to fill out guys' balding manes, many people would buy it simply because they trust Tom Hanks and his "nice guy" persona.*

Actress Jamie Lee Curtis appeared in ads for Activia to sell the public on the digestive merits of the yogurt. Who wouldn't believe the claim if a talented, beautiful, distinguished actress such as Curtis enthusiastically implied that Activia yogurt would facilitate their regularity and be an important part of a "healthy lifestyle"?

---

* I can personally attest to Tom Hanks's genuinely being a nice person. In 1988 I was in college on a date and driving my 1958 Nash Metropolitan. The battery died. The ever-friendly Hanks hollered, "Hey there!" and sprinted across a few front yards into the street and put his hands by the taillight to help me push and "bump-start" my car.

We don't know what occurred with Curtis, who perhaps was in the dark about the false claims made by her benefactors, but to the millions of people who have low PI and don't know which yogurt brand is better than another, her word tipped the scales in favor of buying the product she recommended.

When it comes to the life span of a battery, how many youthful years a skin cream will add to our complexion, or how a supplement will help us shed pounds, it's vital to take a step back and think about whether the product appeals to our low PI through use of celebrity endorsers, socially driven techniques, or claims that seem too good to be true.* While millions of people continue to fall prey to advertising, regardless of its legitimacy, and become seduced by certain products, the repercussions are not nearly as severe as when followers succumb to the pitches and pronouncements of false messiahs and zealots. This contagion can be far more menacing and ruinous to society. We will identify the warning signs in the next chapter.

---

* Well-established brands, like celebrities, can also influence those with low PI because of the, well, branding they have established over the years. My daughters assisted me two years ago in a study assessing the degree of ultraviolet (UV) light blockage in sunglasses. We found excellent UV protection in all the sunglasses we tested — from the cheap to the expensive. We even went down to the Venice Beach Boardwalk armed with my UV meter and were stunned to find that even the inexpensive sunglasses in those shops had terrific UV blockage. Ditto for sunglasses sold at gas stations that we tested. We (myself included before this study) perceive that expensive brands have better UV protection, but in fact, even a $10 pair of sunglasses can have excellent UV protection, thanks to US federal regulations in this area. So go ahead — save some money and purchase a value-priced pair of sunglasses, with confidence (at least in the United States).

13 Fanaticism

The Nature of Extreme Beliefs

It wouldn't be a stretch to declare that we are bombarded by acts of fanaticism and zealotry everywhere we look. We see and hear radical behavior and religion-inspired terrorism around the world with such frequency and immediacy that we are slowly growing accustomed to hearing it — whether it's a mall shooting, terrorists beheading captives, or a suicide bombing in a public marketplace.

Extremism can manifest in many forms, not all of them necessarily involving physical violence or even religion (though this is common when it comes to cult behaviors, as we'll discover). Animal-rights fanatics have been known to splash red paint on the garments of fur-wearing celebrities. Radical antismokers protest against anyone who lights up, even if the smokers stay in their designated areas. Fired-up antiwar demonstrators have conducted *vomit-ins* to protest America's involvement in overseas war. (Yes, this actually occurred in San Francisco.)

Merely having a strong viewpoint, expressing it, defending it, and attempting (within reasonable methods) to inspire or influence others does not necessarily qualify a person as a fanatic. To be sure, I do not presume to judge or categorize any religions, religious leaders, or devout followers. Our goal in this chapter is to address how having low Perceptual Intelligence makes one more susceptible to being vacuumed into fanatical worlds where view-points lack all sense of logic and reason and tip over into immoral and dangerous behavior. And I'm not talking about the cadre of teenage gamers transfixed to their PlayStations — that is literally child's play compared to what we are about to get into.

In earlier chapters we covered lighter examples of fanaticism — such as in the cases of sports fans and groupies, which demonstrate low PI among individuals who prioritize their passions above things that truly matter, such as careers, friendships, and romantic involvements. But when a cause makes one deaf to others' opinions, incites self-righteous behavior, and shuts everything else out, it exposes the lowest possible degree of PI. Unlike shoppers who concede to social influence tactics and maybe buy too much stuff, religious extremists and cult members have such low PI they become vulnerable to brainwashing that risks inflicting irreparable harm on themselves, their families, and society.

Why It Isn't Nuts to Sign Up to Be an Extremist

An individual's thinking is the result of hundreds of thousands of perceptions, judgments, experiences, and biases that marinate over time. When a person is deprived of a stable home during his or her childhood, survives significant emotional trauma, or fails to receive a basic education, he or she becomes a blank (or semibiased) slate with low PI and is thus receptive to inputs from whatever influencing factors happen to stroll by (or these days, infiltrate his or her digital space), eventually being unable to discern reality from fantasy. Either the individual is wandering without direction or has endured so much pain that his or her reality has been severely dampened, enabling insidious ideas to creep in and sink its hooks. People who come from broken, abusive, and/ or unloving homes have gaps in their emotional makeups; they might feel alone, betrayed, cast out, or just different. When these individuals reach low ebb, they might also feel lost and hopeless. Are they crazy, or just looking for a warm hug?

Given the right circumstances and timing (such as just being in a funk or in a clash with family), it's entirely possible for any one of us — including your children — to be caught off guard when our PI is low enough to fall prey to the ingenious mind-altering techniques of devious cult leaders. You are probably shaking your head in disagreement. *How could my children possibly be that gullible?* I'll tell you how: through the leader's unconditional love, the support and camaraderie of a new family, and a new mission or outlook on life.

This is how it generally works: Newcomers to cults are wholly welcomed by the leader and the group, no matter what baggage they might have. Gone are the unrealistic expectations of parents, the mind-numbing pressures of work or school, and the harsh judgments of society. Victims of religious and cult brainwashing are freed from their pasts and given all-new identities; in essence, they are awarded fresh new starts. (Indeed, sometimes they are presented with new names during rites or indoctrination proceedings.) With their new personas in place, they are led into a fantasy world that allows them to experience what it's like to be part of something grand, intriguing, and *important*. Why not join when they have nothing left to lose?

Not only is it untrue that all cult members are crazy, but it's an equal fallacy that they have low intelligence. In fact, several published accounts by individuals who have escaped from cults and overcome brainwashing suggest that the opposite is true. There is the harrowing tale of Elizabeth R. Burchard, who chronicled her experiences as a member of a New Jersey cult in her 2011 book, *The Cult Next Door*. Burchard, an urban teenage girl, was no slouch intellectually; she was her high school valedictorian and a student at Swarthmore. She was brainwashed in a Manhattan psychologist's office while seeking therapy and ultimately became involved in an incestuous cult that involved prophecies of Armageddon. For a person who has been brainwashed, fantasy becomes reality and what might have been considered deviant behavior in the past becomes the new normal. With little ability to distinguish doing wrong from doing right — and encouragement from peers to do the former — antisocial, perverse, and destructive behavior becomes possible. The line is no longer blurry, and cult members become laser focused on completing the assigned mission at all costs. People who exhibit extreme behavior think the ends justify the means; they have such low (or nonexistent) PI that they can't tell the difference between the reality on one end of the moral spectrum and that on the other.

The Pope Who Launched Centuries of War in the Name of God

From 1096 to 1272, the world's most aggressive forces were not children who realized they were duped into buying "sea monkeys." Nor were they animal-rights activists who were opposed to using camels as a mode of transportation in the Middle East. They were European Christians who sought to "reclaim" the Holy Land (Jerusalem) from the Muslim and Turkish peoples. The nine military actions known as the Crusades (one heck of a euphemism) resulted in death tolls somewhere in the range of one to nine million. (Apparently, no one kept good count of such trivialities in those days.)

What sparked such a long-running, barbaric bloodbath? These God-fearing soldiers didn't wake up one morning and say to themselves, "Hey, let's conquer some Holy Land today — we can make a killing selling bottled holy water!" The answer, of course, is God, with a little incendiary prodding from Pope Urban II. On November 27, 1095, the French pope — not exactly the fun-loving sort who toured cities in the pope mobile — delivered an inspirational speech for the ages. He stirred as many as a hundred thousand armed Christians to head toward Jerusalem with the words "God wills it" (sometimes translated as "It is the will of God"). Pope Urban II convinced the soldiers that such a noble conquest (i.e., massacring others) would alleviate their sins, and they were promised material riches as well as all kinds of untold other rewards in heaven.

Did God personally give Pope Urban II an insider tip? Did the Lord specify how many innocent people one must kill in order to call it even for past sins? But wait a minute — isn't *killing* a sin according to both the Ten Commandments and the Bible?

Pope Urban II was a gifted orator who had his own motivations for the initial march to the Holy Land, although at first he was responding to a call for assistance from the Byzantine emperor, Alexios I Komnenos, to defend Christian territory in Anatolia from the Seljuk Turks. The pope's persistent dogma was that Christians should control the Holy Land because it is where Jesus Christ died and was reborn. He was also at least somewhat aware that Muslims had a certain amount of accumulated wealth. When it comes down to it, though, he was a hate-monger who cringed at the thought of heathens who clung to beliefs different from his own. The intent of his words in the name of religion was no different in intent from the speeches later delivered by Adolf Hitler, which were as designed to brainwash the masses by uniting the people in a single cause against a common perceived enemy. The pope referred to the Muslims as "pagans" who "worship demons," while the Christian soldiers were led to believe that God served "as their guide." Once on the warpath, the Christian soldiers were unflappable in their mission, wholeheartedly convinced they had the Lord on their side. Low Perceptual Intelligence among the ranks enabled the sentiment to spread throughout Europe and from generation to generation — even when the rewards received in their lifetimes (we can't speak to what occurred in the hereafter) dwindled.

Ironically, Pope Urban II died before having heard of the eventual capture of Jerusalem (which didn't remain stable for long). He would have no inkling of the numerous subsequent Crusades and the million (or millions) of lives they would claim, all due to the grim enterprise he had initiated.

When Is a Religion Really a Cult?

The Crusades were by no means the only religious war, though it was the one that lasted the longest and is most often depicted in religious art and literature. History is filled with examples of later wars waged against other peoples in the name of religion and/or a higher power, with legions of followers thoroughly convinced by a fire-and-brimstone leader that the cause was justified.

One recent example, the Second Sudanese Civil War, raged between 1983 and 2005 and claimed one to two million lives. The conflict was sparked by President Gaafar Nimeiry's decision to enforce Sharia law on the unwilling Sudanese people — and who could blame them for their resistance? Sharia law involves extreme punishments for crimes and beliefs, such as amputation of hands for stealing and up to a hundred lashes for acts of premarital sex, infidelity, and homosexuality.

In many countries today, such as Afghanistan, a large percentage of the Muslim population continues to believe in Sharia law, albeit with various interpretations and wide ranges of severity. This question is therefore often asked: At what point do radical practices tilt a religion to becoming a cult?

There is a theory that a movement is considered a religion if it has existed for centuries, whereas a cult is the name given to newer groups. Cults often have offshoot philosophies of existing religions but with some subversion in beliefs, rites, and/or practices. As mentioned above, cults tend to have a dynamic, charismatic, beloved leader at the top whose whims drive all decision making. Cults are highly exploitive — for power, control, domination, money, sex, or a combination of all these — and involve various forms of brainwashing, if not total mind control. People who join cults generally sacrifice their sense of self for the whole, whereas most religions allow at least a modicum of individuality outside the main precepts.

In my view, Islam is a religion but *radical Islamic terrorism* — notably ISIS — is cultish, and members have low PI even if they seem to have normal or even high cognitive intelligence. Terrorist groups, Islamic or otherwise (and I certainly don't mean to say that terrorism is solely a Muslim phenomenon), do the following: recruit and brainwash members (usually young), randomly attack innocent people (without any thought that compatriots might be among the mix of those harmed), and willingly sacrifice their own lives based on promised rewards in the afterlife.

Come as You Are: The Blissful Embrace of Cults

Dr. Marvin Galper is a San Diego–based clinical psychologist who worked with cult survivors. With extraction teams who infiltrate the compounds, sometimes under the cloak of darkness, he has helped rescue and debrief (i.e., provide therapy to deprogram cult brainwashing) many young cult members over the years. He described the cult selection process to me: "They seek out people in crisis, such as a college student who failed an examination and has lost family support. They look for kids of broken homes, such as from divorce or loss.... [who] are often alienated from their families." In other words, cults prey on those with emotional instability (which, again, might be just a phase caused by circumstances) that led to low or no PI — that is, no ability to discern whether the suggestions presented to them are true or not.

Galper describes brainwashed cult members as "being in an awful state of consciousness." Cults gradually employ their techniques so the victims are unaware of what is happening to them. Often they are personally welcomed into the group by the leader, who offers "unconditional love" (which is often lacking in their lives) and ample sympathy and support for past suffering. They are accepted "just as they are" by the followers. Over time, they are worked over through an arsenal of manipulative attacks on their minds and bodies, including sensory deprivation (such as light), sleep deprivation, and dietary restrictions. The critical aspect is conformity; cult members are programmed to develop a sense of belonging and oneness with the group, which thwarts a person's individuality and clamps down self-expression. The cults may sometimes incorporate symbols, secret codes, and chants into the repertoire that tap into imagery embedded in the minds of vulnerable people with low PI; this helps create balance, acceptance, legitimacy, and order through vague familiarity with the images. Cults fill a cup that is empty.

Cults may borrow aspects from other religions and sects, but according to Dr. Galper, the leader is "omnipotent and all-powerful.... He has complete access to 'God' and portrays everyone else as being in darkness." Galper sees common traits among cult leaders: they are "charismatic and unscrupulous father figures" with an obsession with "power, domination, and wealth." Many of these leaders take full advantage of their dominant role by forcing their sexual will on women and men alike.

WHAT DO TOM CRUISE, KELLY PRESTON, JOHN TRAVOLTA, KIRSTIE ALLEY, ELISABETH MOSS, ISAAC HAYES, BECK, AND ANNE ARCHER HAVE IN COMMON?

They have all belonged to the Church of Scientology. Is Scientology a religion or a cult? Are the followers spiritual or fanatical? Or are they "scientists"? Why are so many celebrities drawn to Scientology? Do they have low or high PI?

Scientology was the brainchild of L. Ron Hubbard, a sci-fi/fantasy writer. In 1950 he assembled his unique thoughts on spirituality of the mind in the bestselling book *Dianetics*, in which he expounded on the theory that illness was psychosomatic and could be prevented and/or reversed by editing detrimental thoughts out of the brain. But it's much more than that. With the aid of Hubbard's zealous imagination, Scientology states that he was dictator of the "Galactic Confederacy," and 75 million years ago brought billions of people to Earth. I don't think many astrophysicists or anthropologists would concur that these two claims are remotely possible. Billions of human beings were not running around with the dinosaurs, which had become extinct 65 million years ago. Hubbard's concept turned into a phenomenon with millions of devotees — many of whom eagerly gave up their lives for the cause that he created. It's as if George Lucas had established a Star Wars religion based on his blockbuster sci-fi franchise.

Although Scientology has some superficial aspects of religion, I believe it more closely resembles a cult and that its members have low PI. Their sense of reality has been compromised; they believe Scientology holds the power to unlock all kinds of wondrous supernatural gifts from within. If you are interested in tapping into the "power of positive thinking," I recommend that you read the works of the late Norman Vincent Peale, whose statements are more grounded and whose books are a heck of a lot less expensive than joining the Church of Scientology.

Several well-known cult populations have met with tragic ends via mass suicides. When people with low PIs are brainwashed to such an extraordinary degree, they are capable of believing and doing, well, just about anything suggested by their cult leaders, no matter how preposterous. In 1978 Jim Jones, founder of the Peoples Temple cult, directed *918 people* — including 276 children — to imbibe a fatal drink resembling grape Kool-Aid that was laced with cyanide. Nearly twenty years later, in 1997, the Heaven's Gate cult (created by Marshall Applewhite and Bonnie Nettles) coaxed thirty-nine cult members to swallow a deathly tincture that claimed their lives. Why? Because they believed they would somehow climb aboard a spaceship that trailed the Hale-Bopp Comet. Believe it or not, this cult still exists, with a few hundred remnant followers still waiting for transporters to energize.

.Cults come in a variety of forms and may be orchestrated by people with all kinds of narcissistic tendencies, including faith healers, televangelists, and preachers. Representatives might appear at one's front door or infiltrate a person's life through solicitation via the Internet, email spam, or old-school snail mail. If you know of a person under twenty-one who has been cast out by his or her family, you may want to pay close attention to whether he or she is exhibiting any unusual behavior (incessantly singing Ariana Grande and Nicki Minaj songs excluded), such as verbalizing strange-sounding beliefs, wearing unusual clothing, expressing a dislike of past interests, possessing strange pamphlets, and demonstrating a lack of ability to think critically. If you happen to suspect someone is trying to lure you or someone you know into a cult, you can turn to various available anticult hotlines, help groups and, of course, the authorities.

While some cult leaders are convinced they are gods who will endure forever, time has proved otherwise. (I'm still waiting for L. Ron Hubbard to reappear with just a few people; I'm not even holding him to a million or billion.) As stated earlier, it is possible for some cult survivors to be debriefed and for the brainwashing to be cleansed, though Dr. Galper cautions: "It takes a strong support system to overcome losing the union with the cult community. When [former cult members] return to regular life they have to find another community." The strongest support system comes as family and close friends.

In time cult members can transition back into the mainstream once they fully recognize that their cult educations were a sham and have started being able to discern fact from fiction — an act that improves their Perceptual Intelligence. As it happens, *time* — the focus of our next chapter — is a key element of PI in terms of how we interpret its passage, how it distorts our memories, and how it can change, based on one's experiences. To quote Mother Teresa: "Yesterday is gone. Tomorrow has not yet come. We have only today. Let us begin."

14

The Subjective Experience of Time

And a Chapter Bonus: The Origin of the Bucket List

It's Sunday night. Jonathan, a pimply twelve-year-old boy, becomes frantic as he realizes that he still needs to write an essay that's due the next day. He's wasted the entire weekend watching all the Harry Potter movies twice, and now he has to hunker down and get his assignment done. He plunks himself at his desk, turns on his computer, and mumbles to himself: "Why'd I wait so long? I hate Sunday nights; they come so freakin' fast."

Elsewhere in the same house, situated in the home office, Jonathan's father fires up his own laptop. He must deliver a major presentation to his boss the next morning and hasn't looked at the PowerPoint deck in more than a week. He thought he had plenty of time to prepare, but, well, between carpooling his son to and from basketball and soccer and watching basketball on TV with his face-painted buddies and a lot of beer, he lost track of time. He receives an email from his boss saying that his presentation has been moved up to an hour earlier. He has a near panic attack as he waits for his presentation to load. "Damn, why did I wait so long to prepare?" he grumbles to himself. "The weekend just flew by."

The apple doesn't fall far from the procrastination tree

as father and son diligently scramble at their respective laptops. Suddenly, the doorbell rings. Grandpa, who's eighty, has decided to stop by. A couple of moments later, Mom announces to the household: "Honey! Jonathan! Grandpa's here! Come on down!"

Father and son stumble toward Grandpa, who notices things aren't quite right. "What's wrong, boys? Did I come at a bad time?"

"It's not you," his son-in-law assures him. "I have a big presentation early tomorrow."

"And...I gotta finish an essay for school," Jonathan says.

"I'm so sorry. I didn't know," Grandpa says.

"It's Sunday night, Dad," his daughter points out. "You remember what that felt like, don't you? Time just slips through our fingers, and the weekend is suddenly over."

"It's Sunday night? Golly! The weekend just came and went! I have to go — I'm late. I have my second dinner date with Gloria — she's got such a beautiful smile. Bye, everyone!"

The above story isn't designed just to present three generations who have little sense of time. Its purpose is to demonstrate low PI when it comes to *interpretations* of time:

- The son who procrastinated on his assignment and believes that *Sunday nights come so freakin' fast.*
- The father who was so distracted by chores and the NBA that he didn't prepare for his presentation and is convinced that *the weekend just flew by.*
- The mother who reminds Grandpa what time was like as it *slipped through our fingers, and the weekend is suddenly over.*
- The grandfather who doesn't pay attention to what day of the week it is and thinks the weekend *came and went.*

The scenario played itself out for all four parties during the same weekend, in the same house, and yet everyone chose different words to describe how quickly they felt time had elapsed. Unless the family took a detour with Rod Serling in *The Twilight Zone*, a weekend is always going to be the same forty-eight hours for everyone involved.

In a 2016 study published by the *Journal of Scientific Reports*, researchers found that the human brain more often than not perceives time incorrectly. Through a series of tests in which participants were asked to identify the time spans between various visuals and sounds administered in patterns, the subjects tended to base time estimates somewhere between reality and their perceptions of *what they thought the time frame should be and what would occur next*. This phenomenon comes with pros and cons for human beings; we are inaccurate in assessing time frames, but, when we get ahead of ourselves in this way, our perceptions help prepare us for future events. As Dr. Max Di Luca of the University of Birmingham said, "These predictions are essential to survival because they allow us to react faster to the environment and plan what actions to perform." High PI sometimes means not caring about time precision when we are anticipating something that benefits us or that involves self-preservation.

As we age, time seems to pass much more rapidly than when we are younger. Researchers at the San Jose Faculty of Medicine in Brazil found that all age groups (between fifteen and eighty-nine) tested in a 2016 study mischaracterized the passage of time by judging it too quickly — but the oldest individuals fared the worst. There isn't a conclusive reason for this, but the scientists speculate that changes in dopamine occurring as we age might interfere with memory and concentration, which then alters our perception of time. There is also the potential psychological impact of aging and the need to "get things done while there is still time left" which is why people create "bucket lists." (The term is connected to "kicking the bucket," a phrase that comes from hanging oneself and then literally kicking the bucket out from under.)

The Fabric of Time: What Would You Believe in a Different Era?

When it comes to understanding time, our Perceptual Intelligence is a dynamic ability. I have a theory that, as social perceptions change and evolve (when was the last time you handwrote a thank-you card?), so does our collective thinking. Centuries ago people thought the world was flat and the moon was made of cheese. (Swiss, if I'm not mistaken.) Thanks to Buzz Aldrin and Neil Armstrong, we now know with complete certainty that the moon is *not* made of cheese (money well spent by NASA).

In spite of all the science confirming climate change and evolution, deniers of these incontrovertible facts abound, some of whom are well educated and hold prominent positions. Their stubbornness in refuting science of such conclusive magnitude is due to two root causes: (1) The issues fly in the face of their financial, political, or religious interests and/or long-held beliefs, and (2) the individuals need to see some concrete evidence, like Florida sinking into the ocean, in order to accept climate change as real. Many people seemingly have no control over their beliefs because their biases have locked their opinions so firmly into place. With regard to evolution, it's unlikely that acceptance will occur for these doubters unless scientists can clone or reanimate a caveman (or we devolve and become monkeys). Climate change is a far dicier matter when it comes to the acceptance of science because in many cases financial interests have clouded opinions, resulting in low PI in that area. For some it will only become a reality a hundred years from now (or however long it takes) when the ice caps have melted and we all live on boats.

I believe that time has the ability to change our perceptions and opinions about pretty much everything. In chapter 10 we discussed how Oscar Wilde was persecuted for his homosexuality; although discrimination still exists, and more in some parts of the country than in others, on the whole society is a lot more understanding about sexual orientation, and people in all arenas have openly come out of the closet without repercussion. A talk show hosted by an openly lesbian comedian like Ellen DeGeneres would not have been possible back in 1960, 1970, or even 1980, but today she is a popular and beloved figure with strong TV ratings and career longevity.

Life experiences can have a significant impact on whether we have low or high PI. Many of us know at least one person who was a lifelong homophobe but turned the corner when he attended his son's gay wedding ceremony and embraced his son-in-law as family, or some story along those lines. Even bigoted, homophobic TV character Archie Bunker (from the groundbreaking 1970s TV show *All in the Family*) mellowed out in his later years as the times changed (although he never failed to call his liberal son-in-law, Mike, Meathead).

Does Anyone Really Know What Time It Is?

We have radically different perceptions of time based on a wide range of factors, including our age, our location, and our personal short- or long-term situation. A toddler will not view time in the same way as her mother pushing her in a stroller through Target because, according to child development psychologist Penelope Leach, her memory has not been developed and therefore "she can't wait a second for anything." A woman in a seemingly never-ending line at the Department of Motor Vehicles will view her wait as being far longer than if she were spending that same amount of time having fun at Disney World. A corporate CEO might label a restaurant's service as slow because it took ten minutes to bring out his or her drink and appetizer; by contrast, a young romantic couple at this restaurant might happily receive their drinks and appetizer within the same time frame and later post a favorable review online about the establishment as having "lightning-fast service." The CEO and the young couple have completely different perspectives on the service because of their mind-sets at the time. The CEO might be stressed out because of a business deal going south and because he or she is late for another meeting. The couple, by contrast, might have all the time in the world because they are enjoying themselves and want to stretch out the moment.

It's about Time for the Time Clichés to Stop

There are probably more clichés about time than any other subject, except perhaps love. Chances are you've heard one or more of these within the past week or used one yourself without even thinking about it:

- Time stood still.
- A stitch in time saves nine.
- All in due time.
- At the last minute.
- Hold on a second.
- Hold on a minute.
- Rome wasn't built in a day.
- It feels like it was only yesterday.
- It felt like a lifetime ago.
- It felt like another lifetime.
- Better late than never.
- Just wait a gosh-darn minute.
- Hurry up; I'm growing old from waiting.
- Time heals all wounds.
- Time is on my side.
- Time waits for no one.

Do the above sound familiar? (The last two should; they are, after all, also the names of Rolling Stones songs, albeit with completely opposite messages.) Do they have a ring of truth? Probably not — at least in raw form, as there isn't any context to go along with them. In order for human beings to demonstrate their

Perceptual Intelligence regarding time, they must have surrounding details to help assemble thoughts and draw conclusions.

Animals have completely different perceptions of time than human beings. When we attempt to swat a fly, we must seem pretty lame to the fly because, according to *Scientific American*, these insects have the ability to process four times more visual information per second than we do. Time perception for animals is largely based on their size, metabolic rate, and environment. The smaller the creature, the slower everything seems to them. Many domesticated dogs, according to Animal Planet, are so emotional and/or anxiety ridden (yes, doggie Valium now exists) that any kind of separation from their beloved owners can discombobulate their perceptions of time. You could leave your house for five minutes or five hours, and your dog will still be waiting for you by the front door as if you've been gone a lifetime.

Among all the many factors affecting us, our ages especially influence our sense of time and the accuracy of our perceptions. If you were to ask my young daughters to define *old* they would probably say, "Well, *you're* really old, Daddy." Ouch! I only recently turned fifty. Is that old? By today's standards fifty isn't old at all, but to a young child it might seem that way in comparison to her own age. Her PI with regards to age hasn't yet formed. But imagine if I lived in 1907, when the average male life span was 45.6 years — I'd be considered an old geezer who had already outlived his time. And, if you go back to classical Greece, life spans were only in the twenty-five to twenty-eight range. I would be a *fossil* (or at least on my way to becoming fossilized) by those standards.

Cultural influences also impact how we perceive our time on Earth. As medical advancements and quality of life improve at rapid rates, life spans continue to grow, and our perceptions of age shift as well. If you're lucky enough to live in Monaco, the average life span is a ripe 89.52. But in other parts of the world, life spans are declining due to poor medical care and harsh climates; in Chad, Africa, for example, the average life span is only 49.81. "Old" in Chad is very different from "old" in Monaco.

In recent years we've heard expressions such as "sixty is the new fifty" and "fifty is the new forty" to give ourselves psychological edges in terms of how we perceive our ages. All this is reasonable, given the prevalence of skin creams, special diets, elaborate exercise regimes, and cosmetic surgery. However, it is still a sure sign of significantly low PI for someone to act (or sometimes dress) as if he were a nineteen-year-old athlete when in reality he is eighty-nine and in pretty bad shape.

The Fault Lies in Our Memories, Not in San Andreas

Human memory is a tricky thing. Our recall of an event never precisely duplicates the event itself. Memory is an ever-evolving act of our creation. Our ages, life events, and even dreams distort our memories and interpretation of time, meaning that our perceptions of time will always be subjective. How do we know when we have low or high PI when it comes to time if everyone is viewing it through a different lens? Consider the following scenarios and how you would regard the following activities during a twenty-minute span:

- Making love with your significant other
- Eating an ice cream sundae
- Waiting for the season finale of your favorite show to start
- Being grilled by your boss
- Lying on your deathbed

Twenty minutes would feel very different to you depending on which of the above situations you were in, and the degree of difference would vary greatly from person to person. If you like and respect your boss and think she is making a fair point while grilling you, maybe the twenty minutes is a learning experience and doesn't feel too long. If, however, you dislike or fear your boss, and she is haranguing you, twenty minutes could feel like a century. And yet all these occurrences are happening in the same amount of time.

Time itself may be unalterable, but our perceptions of it certainly change. We human beings even have the power to shape our perceptions of time and benefit from this experience, something no other species is capable of. Many yogis and spiritual gurus spend their entire lifetimes attempting to "live in the moment," to slow down their perceptions and connect more with their inner selves and fully embrace the world around them. But even the most revered yogi is still beholden to the realities of time, just like the rest of us.

PHIL SAYS IT'S GOING TO BE A LONG WINTER

Most of us likely remember *Groundhog Day*, the hilarious 1993 film starring Bill Murray as a weatherman named Phil who is doomed to replay the same day, Groundhog Day, until he figures out how to get it right. During his interminable experience in Punxsutawney, Pennsylvania, with the forces of déjà vu set upon him, Phil suffers all kinds of highs and lows: he repeatedly steps in the same puddle, he saves lives, while suffering an existential depression he takes his own life in all kinds of crazy ways (electrocution in the bathtub, jumping off a building, and stealing the town's famous groundhog, who also happens to be named Phil, and then driving off a cliff with him), studies the beautiful women in order to seduce them, becomes a master pianist, creates ice sculptures, and on and on — all in the same repeated day!

Phil's Perceptual Intelligence regarding time becomes so godlike that, by the final sequence of the film (which I won't give away, on the rare chance you haven't seen the movie, which has also been turned into a Broadway musical), he figures out with precision what he plans to do with every split second of the day to make it perfect. He accepts his fate in this self-repeating fantasy world and turns his impeccably accurate perceptions of the towns-folk in his favor, though no longer with malicious intent. One could say he has become mindful and respectful of time, learning to accept his circumstances and severing all his past views to reach a higher truth about himself and everyone around him.

Is it possible to slow down or speed up our perceptions of time? According to a 2015 study, the answer is yes. Researcher Aoife McLoughlin of James Cook University attributes most of our current incorrect perceptions of time to technology. Our smartphones and laptops help us do many things leagues faster than ever before — research, communication, and all manner of business — but we are being tricked into thinking that time is moving faster than it really is. "While it might help us to work faster," McLoughlin says, "it also makes us feel more pressured by time." This discovery was made simply by comparing people who used technology to those who didn't.

The simple solution? Unplug. Shut it down. Take an electronics break. If you can limit phone and Internet time at work to two calendar appointments a day of maybe a half hour each, you will find yourself less distracted by all the noise. Time will seem to slow down, and you will find more time to relax or to accomplish the things on your to-do list. This is even more beneficial on weekends and vacations: Why waste precious leisure time on electronics when you can be out and about doing things, having fun, and making time seem to pass more slowly?

However, if you are waiting in line at the Department of Motor Vehicles or in the waiting room at jury duty, assuming that electronics are allowed in those situations, then by all means go online and see if you can speed up a time period that otherwise feels draggy and unproductive.

As this chapter has demonstrated, our perception of time isn't fixed: it bends, entertains, and baffles us. It's shaped in our minds by our age, our culture, and our wills and, in turn, determines our destinies. We slow it down and speed it up with alterations in our PI. We can use it to hold on to the past or exist more productively in the present. It is, in fact, our perception of time that separates us from the apes, the dogs, the flies, and even the groundhogs. At the same time, it is our intuition — whether we choose to pay attention to it or ignore it — that causes us to succeed or fail in our life's quests with what little time we have on Earth. Let's take a look at that now.

15 Gut Check

Following Our Intuition

Your cell phone rings. You don't know why, but you have an eerie feeling that something terrible has happened. A sudden thought strikes you, and a chill runs up your spine: your beloved aunt has died. She was healthy when you saw her a month earlier, and you've been so busy you haven't thought of her since. Why did that macabre thought materialize out of the blue at that precise moment? Did the ring tone "sound different"? No, of course not. You answer the call. It's your mom telling you what you already knew. But did you really know? How is that possible?

Some people refer to the bizarre feeling of knowing in advance that something, good or bad, is about to happen as a "hunch" or a "gut feeling." Typically, we can't explain or describe it when it strikes us; we simply choose to heed an intuitive signal or to ignore it and then regret it later. Our ability to recognize fleeting gut sensations and to make the correct decisions about them at the right moment demonstrates whether we have low or high Perceptual Intelligence.

Relax: I'm not going to hype new age theories about supernatural forces at work to explain this phenomenon. Rather, I'm looking to investigate a phenomenon, known as *intuition*, that we have all experienced to some degree. This is another sense in our arsenal of natural abilities, though we hardly comprehend it and barely have a clue how to put it to good use — presenting a great opportunity to improve our PI.

Seldom is a gut feeling grounded in logic or reason. It can appear as an idea that floats into your head and vanishes as quickly as it arrived. It can also appear in a dream or as a hallucination and may concern an event that affects you or other people. When we opt to follow our inner voices and pounce, despite the absence of data, when we "trust our gut," more often than not we find ourselves guessing correctly. Depending on the situation, these "guesses" sometimes mean the difference between failing and succeeding or even living and dying.

I view intuition in two ways: as non-science-based, such as when astonishing ideas come "out of the blue" or in a dream, and science-based, such as those intuitive skills that have been studied and proved to exist, at least to some degree. In this chapter, we'll address the non-science-based kind first — since, frankly, it's more fun.

The Day the Music Died — and Intuition Failed

On February 3, 1959, rock stars Buddy Holly, the Big Bopper (J.P. Richardson), and Ritchie Valens perished in a plane crash in Iowa while on a lengthy musical tour. The event has become part of rock-and-roll lore, as documented by the "day the music died" lyric in Don McLean's classic song "American Pie." What is not quite as well known is that three other musicians might have been on that plane had it not been for some pretty odd circumstances: singer/songwriter Dion DiMucci (known best as part of Dion and the Belmonts) did not get on the plane because he didn't wish to pay the $36 fee; rockabilly performer Tommy Allsup wanted to fly but lost a coin toss to Ritchie Valens; and singer/songwriter Waylon Jennings (then Holly's bassist) courteously gave up his plane seat to the Big Bopper, who was ill with the flu. Years later

Jennings recounted in his autobiography that, in anticipation of being relegated to riding on a freezing bus instead of boarding the plane, he jested to Buddy Holly: "I hope your damn plane crashes!" Jennings's parting words turned tragic: the plane went down in a cornfield shortly after takeoff, claiming the lives of the pilot and the three musicians.

What intrigues me most about this story is that documentation shows that each of the three artists who died — Holly, the Big Bopper, and Valens — had experienced premonitions of their deaths, and yet *they were the ones who ended up on the plane*. The backstory in each case is chilling — even if you are a skeptic:

- *Buddy Holly*: Maria Elena Holly, Buddy's widow, revealed that she'd had a nightmare of the plane crash not long before it occurred. On awakening, she described the catastrophe to her husband who, in turn, admitted that he'd had a similar dream in which he saw himself dying in an airplane crash on a farm. As if this isn't creepy enough, in 1958 a producer named Joe Meek happened to throw some tarot cards that conveyed the words "Buddy Holly" and "dies," with the date February 3. Meek warned Holly in a letter and, on that date, an autograph book attached to a brick (believed to have been tossed by a fan) sailed through Holly's dressing room window. Did he survive? Yes and no. He evaded the brick hurtling toward his head on February 3, 1958, but the fatal plane crash occurred *one year to the day later*, on February 3, 1959.

- *Ritchie Valens*: Although he was only a teenager, Valens had a well-known fear of planes and vowed he would never fly. Most likely his phobia began as a result of having witnessed firsthand a plane crash that had claimed his grandfather's life.

- *The Big Bopper*: After being awake for three days while serving as a DJ for a Disc-A-Thon, the Bopper suffered from severe exhaustion and reported seeing hallucinations —

including one of his own death. He is reported to have said, "The other side wasn't that bad."

Could these deaths have been averted if the musicians had heeded these intuitive signs? We'll never know, of course, but it should give us some pause; when we receive those inexplicable "signs" ourselves, we should at least evaluate them before doing the equivalent of boarding the plane.

Intuition: It Can Take You for a Ride

Intuitive signals don't always concern being informed of a beloved family member's death — or dreaming about your own. Sometimes they're as innocuous as thinking about a song and then hearing it right when you turn on the radio. And there are successful people who seem to have a knack for capturing and exploiting "the right thing at the right time." The greatest minds rely heavily on their intuition — not just talent or intellect — whether it was Albert Einstein imagining chasing a beam of light when he was just sixteen, Thomas Edison inventing the phonograph, Nikola Tesla developing the alternating-current electrical system, the Beatles noodling with their next studio innovation, or even Oprah Winfrey placing her finger on the pulse of the nation's major social issues and interests. The most brilliant ideas and breakthroughs are often said to just "come" to these individuals, after which they exploit them, using their skills, experience, talents, and intellects.

By no means is this process limited just to inventors and creative artists. The most successful business professionals and investors also have an innate ability to capture flashes appearing in their minds that at first seem counterintuitive but become game changers once acted on; the accomplishments of these folks make them seem godlike to everyone looking in from the outside. The CFA — an organization devoted to education and ethics in the investment community — is a staunch supporter of using intuition to help make investing decisions. They cite George Soros and quant authority Emanuel Derman as two giants who have done so with tremendous results, and CFA's *Enterprising Investor* columnist Jason Voss offered an article, "The Intuitive Investor," in which he presents a compelling case for gut feeling when it comes to money matters: "Over the course of my investment career, I used several unconventional tools to improve the results of the fund I co-managed, but none was more powerful than intuition."

Always Wear Your Safety Belt, and Don't Forget the Toilet Paper

When I was on the college debate team at UCLA, on occasion we had to drive great distances to tournaments. On one trip, while our team was returning from a match in Paso Robles, California (about four hours from Los Angeles), we traveled in a beat-up station wagon donated by a debate alumnus. I always wore a seat belt, but I was in the middle seat in the back row of the station wagon and the seat belt was wedged into the crack of the seat, so I decided not to yank it out and put it on this time. I was seat belt–free as we sped along the freeway. About thirty minutes into the drive, I was hit by an unexpected urge to dig out the seat belt and put it on. I didn't think much about this gut feeling as I excavated the seat-belt straps and clicked them around my waist. No more than thirty seconds later, I heard a loud pop from a blown tire. The station wagon spun out of control and off the highway, landing in a ditch. Fortunately, everyone had been wearing a seat belt and no one was hurt. Had I not been wearing one, at the least I would have been thrashed about inside the car like loose change spinning around in a dryer. At worst, I would have been ejected through the open window. This is an example of how high PI — listening to the inner voice without question in the moment — can have a potentially lifesaving impact.

By contrast, I've also observed the consequences of low PI when someone close to me failed to listen to her gut feeling. A few years ago, my family and I were leaving for a safari in Kenya, when a random idea occurred to my wife, Selina: she sensed she should bring toilet paper. She ignored the hunch. We flew to Migori Airport to go through customs and then transfer for another flight to Tanzania. While at the airport, one of my daughters had to use the bathroom. As it turns out, the toilets there were not quite what we were accustomed to; they were porcelain holes in the floor with ribbed edges for your shoes to grip as you squatted over the hole. (Having strong quads is always a plus for being able to manage these situations.) And guess what? *There was no toilet paper.* The situation seemed hopeless for my poor daughter. Unless one is the child of Superman, it is impossible for a young girl (or boy) to withstand the mounting intestinal pressure that feels like the Hoover Dam holding back a deluge. With no other options, my daughter did what she had to do without the aid of toilet paper. In retrospect, had my wife listened to her gut feeling and brought the roll of toilet paper, my daughter would have stepped outside the bathroom with a face reddened from the high heat but not from embarrassment. Selina had the intuition but, alas, not the PI to make the correct decision in the moment.*

Getting Our "Spidey Senses" to Tingle

We've all had a moment when we predicted that a certain event or thing would happen but could find no real explanation as to how or why. Even animals have been known to have inexplicable intuitive "super senses" and demonstrate abilities we can't begin to fathom: butterflies, and other creatures who use the Earth's electromagnetic field to migrate to the same spot every year; wildlife that flees long before there are any signs of an earthquake or a volcanic eruption; and dogs with schnozzes so powerful they have been known to sniff out cancer. In the fantasy world of comic books and films, no superhero with "animal intuition" is better

---

* My wife, Selina, has excelled in plenty of other situations that demonstrated high PI. (This footnote ensures that I don't get relegated to the couch when this book hits the shelves.)

known than Spider-Man — the arachnid-bitten superhero who anticipates danger with his *spidey sense.*

Spider-Man aside, this second part of our investigation into the world of intuition is science-based, meaning that plenty of research helps us understand why certain people seem naturally gifted when it comes to recognizing intuitive sensations, whereas others miss them completely and just don't "get it." For example, you probably have a friend or family member who has experienced a lifelong history of "bad luck." If you peel back his or her situations over the years, you'll probably arrive at the conclusion that your friend is a chronic overthinker, a person who swats at all the options ad nauseam, changes his or her mind more regularly than Tom Brady tosses touchdown passes, and inevitably makes the wrong decisions after all that wasteful deliberation. The issues might range from choosing which sweater to wear in the morning to making major life decisions, such as choosing a spouse, buying a car, or applying for a job. The "bad luck" comes into play when — after an interminable amount of mind changing — the spouse turns out to be a louse, the car is a lemon, and the job implodes because the company goes under a month later.

People who experience such devastating misses aren't less intelligent than the rest of us, but rather they often fail to notice the signals, don't recognize them for what they are, mistrust them, or can't commit to them. All these scenarios lead to fatal second-guessing. Numerous scientific studies have shown that overanalysis and failing to pay attention to one's first intuitive feeling greatly reduce one's chance of being correct.

In the case of the overthinking friend, you and every other friend and relative (perhaps a few psychologists as well) have probably imparted a lifetime's worth of advice without seeing any progress in your friend's decision-making process. You see her continuing to miss opportunities, and it frustrates and upsets you. You want to help, but what can you do when you know she'll change her mind again in five minutes? I have three suggestions that might help her start listening to her inner voice and heeding her intuition: (1) Take a reputable mindfulness course, (2) find a life coach, and (3) make a list of all her intuitive failings (a.k.a. regrets) on a sheet of paper and say them aloud three times; hearing these spoken aloud might open up her mind to snagging the next opportunity before it floats away.

Winning without Having a Clue How

A pioneer in the field of intuition, Portuguese neuroscientist Dr. Antonio R. Damasio of the University of Southern California published his findings in the *Journal of Neuroscience*, concluding that our gut feelings are, in fact, linked to our brains through emotional memories. His study involved sixteen participants who were "gambling" with four decks of playing cards and $2,000 of play money. Some cards were worth $50 or $100, while other cards lost money. The subjects were asked to flip cards from the decks by their own choosing. Players did not have a clue that the decks were stacked; two were good and two bad, with riskier payouts and greater losses for the latter. The results showed that the subjects with what the scientists deemed normal brain function figured out how to win, even though they had no conscious idea why.

Scientists Galang Lufityanto, Chris Donkin, and Joel Pearson from the University of New South Wales refer to the gut-brain relationship as *nonconscious emotional information*. In their 2016 study, the researchers sought to make intuitive decision-making tangible through a series of experiments with college students (without whom there'd hardly be any studies). The subjects, who were presented with moving visual stimuli resembling snowy television static, were asked to identify the direction of the shifting clouds (left or right). While the clouds of dots were moving, various photographs were superimposed over them that were intended to provide subliminal emotional hints to the participants. In most cases, when the images appeared the students were able to better judge the path of the dots. "Another interesting finding in this study," said Pearson, "is that intuition improved over time."

The results of this study are so striking that they have led some in the scientific community to believe that at some point intuitive ability will be *measurable* — and perhaps even *taught*. Imagine the impact this will have on the following scenarios: students taking multiple-choice tests, compulsive gamblers at casinos, coaches recruiting athletes, and employers hiring creative talent.

In his groundbreaking book *Blink*, Malcolm Gladwell uses the example of the TV show *All in the Family* to demonstrate the power of intuition. All the traditional indicators, such as audience focus groups, showed that the show was guaranteed to be a complete miss. And yet creator Norman Lear and one network executive just knew that they were capturing something fresh, exciting, and revolutionary that had never before been portrayed on prime-time television. In this case, Lear and the TV exec's *expertise* was behind their intuition that they should do the show. There is science behind the concept that *experience breeds intuition*. Gladwell discusses an art historian who can instantly recognize a fake artifact, a tennis coach who almost always predicts when a tennis serve will be a double fault before it occurs, and a lobbyist who had an uncanny ability to identify intangibles indicating whether someone was eventually going to be president of the United States. Individuals who exhibit such unusual skills don't have "psychic abilities" but are subconsciously using their expertise, which produces their "gut feeling." The same cannot be said for the vast number of record companies who turned down the Beatles before producer Sir George Martin, who saw *something* raw in them outside their easily discernible humor and charm, although at first even he had difficulty explaining what it was.

Turning On and Tuning In: Jump-Starting Our Intuition

How do the rest of us learn to harness our gut feelings and act on them to our benefit? How do we become the next Steve Jobs? How do we prevent missing signs on when to avoid plane travel?

How do we recognize if we have low or high PI with regard to intuition?

The key is being *aware* and *accepting* of ideas that strike us (and others) for no reason, even if they seem crazy or go against the grain. Often ideas come to us as we are about to doze off at night or when we start to wake up. In a nanosecond you need to grab the intuition, assess it, and make a real choice. Think about it. Did Picasso spend months deliberating whether he should experiment with cubism? Did Bob Dylan crowdsource the idea of going electric in July 1965?

Sometimes going with your gut means being able to perceive things based on the slightest of details, such as catching the intonation of a colleague's voice or detecting a subtle nuance in her facial expression. On other occasions, it means whisking things out of thin air that magically come into your head — and then having enough conviction to make them happen. Even if you were to place yourself in the company of Einstein and Tesla, be warned that some colleagues, friends, and family may take issue with you following a path based on intuition. Going with your gut often requires a stomach made of steel to counter dissenting forces and negativity.

For all the many artists, scientists, and business professionals who claim to be "struck by genius" while doing their work, there are just as many who find creative inspiration by taking a walk around the block, making a cup of coffee, jogging ten miles on the treadmill, sitting on the "throne," or listening to a track by their favorite musician. Ironically, for those wired a certain way, distraction from the specified activity — doing something completely unrelated to the work — is what provides them with the best ideas. For the late John Lennon, seeing a drawing by his young son Sean led to the masterpiece "Lucy in the Sky with Diamonds" (although drugs and the works of Lewis Carroll had *something* to do with it), and many of his other songs (such as "A Day in the Life") wouldn't have been written if certain newspaper headlines hadn't happened to catch his eye and spark his imagination.

Now that we have addressed important issues such as how to recognize manipulative social influence, dangerous cult behaviors, false sensibilities regarding time, and the dangers of ignoring your intuition, let's put it all together. In the next and last chapter, I provide a quiz to help you assess and develop high Perceptual Intelligence.

16 Your PI Assessment

Can Perceptual Intelligence Be Improved?

As we've seen, it's possible to harness the power of your perceptions, live more consciously, and, ultimately, improve your ability to see, and benefit from, the reality of what's in front of you. I think we all can do a bit better in these areas, and by doing so we can add more joy to our lives.

In the previous chapters, we've seen a pretty comprehensive picture of many seen and unseen assaults against our PIs:

1. How your mind-set can negatively affect your health: hypochondria.
2. Waking up and believing you were just attacked: dark lucid dreams.
3. The narcissism and facades of people in power: Putin.
4. Sports fans taking their love of the game too far: those who riot (publicly or privately) after a championship win (or loss).
5. Individuals believing they see miracles in the mundane: the $28K grilled cheese sandwich.
6. Consumers getting talked into purchasing things they had no intention of buying because of reciprocal sales techniques: car dealers.

7. Purchasing food labeled organic even though it isn't necessarily healthier: the halo effect.
8. Consumers allowing celebrity endorsements to influence their buying decisions: Jamie Lee Curtis.
9. Judging and sometimes even punishing others for having normal biological urges: masturbation.
10. Patrons falling for overpriced and/or rare objects of desire: cat poop coffee for $100 a cup.
11. Making bad financial decisions by doing what your group of friends do: following the herd.
12. Followers being lured in by false ideology: cults.
13. Denying facts due to self-interests and the consensus of the time period: climate change denial.
14. Victims missing intuitive signs that could have benefited, if not spared, their lives: Buddy Holly, the Big Bopper, Ritchie Valens.

As this list reveals, there are so many ways for low Perceptual Intelligence to be a distorting, if not blinding, factor in our lives that it would be impossible for me to capture every possible area of impact (though there is always room for a sequel). For now, to help you figure out how well you discern reality from fantasy and rate on the PI scale, I've provided the following PI Assessment. It has been designed to be fun, quick, and easy. As you answer, go with your honest, first gut instinct response on how you would really react in each situation. Once you've added up your score and determined your result (the scoring key follows — but no peeking!), you'll be more mindful of how you can make decisions, focus your creative energy, let go of inadvertent biases, dismiss fakes, avoid getting rooked, and place your trust in the right people.

PERCEPTUAL INTELLIGENCE ASSESSMENT

Circle one answer for each of the following twenty statements.

1. While sitting in a Starbucks enjoying your coffee, a beautiful woman or a handsome guy hands you some pamphlets and tells you that you have an opportunity to achieve bliss by attending a local gathering run by her or his scholar friend, whom she or he dubs a "genius." You:

   A) Strike up a conversation, review the pamphlets, and attend the gathering out of curiosity.
   B) Ask the person out on a date.
   C) Politely say, "No thank you, I don't subscribe to your religion."
   D) Tell this person to bugger off and call a cop.

2. Your favorite NFL team just made it to the Super Bowl for the first time in twenty years — but they lose the big game in triple overtime on a careless fumble. You:

   A) Sob for three days over a vat of mint chocolate chip ice cream.
   B) Get drunk with your friends in the middle of the night, kick your dog, and start lighting cars on fire.
   C) Fold up your team pajamas for the next few months and start thinking about baseball.
   D) Spend $1,000 on football championship game jerseys and memorabilia.

3. A local politician you support makes a speech on TV saying that the news reports of tainted water are completely untrue and that your drinking water is perfectly safe. As you pour yourself a glass from your tap, you notice a funny smell. You:

A) Drink the water anyway.

B) On second thought, you think you may have imagined the funny smell so you spill out the water, refill the glass from the tap, and guzzle it down.

C) Write to the politician telling him he's a jerk and that you'll never vote for him again.

D) Spill out the water and get it tested before drinking it again.

4. Tom Hanks becomes the product spokesperson for a national life insurance policy. You:

A) Dump your current policy in favor of the one Tom Hanks recommends.

B) Invest in the insurance company Tom Hanks recommends because you sense the stock is going to soar.

C) Verbally trash Tom Hanks's films because he sold out and is advertising products.

D) Spend hours googling the insurance company to find out what they have to offer and why someone as reputable as Tom Hanks would choose to serve as its product spokesperson.

5. The night before taking a plane trip to Brussels, you have a dream in which your aircraft crashes. You wake up in a cold sweat. Right before you share your nightmare with your significant other, he tells you that he had a dream in which you died in a plane crash over Brussels. You:

A) Swear off flying ever again and only travel via ocean liners to Europe, climbing up the ship's bow, putting your arms in the air, and yelling, "I'm king of the world!"

B) Book a different flight.

C) Get boozed up before boarding your plane.

D) Ask all your friends and family what you should do,
    and then go with the majority of responses.

6. You are making your first chicken pot pie. When the
   steaming pie comes out of the oven, you notice the cracks
   in the crust have formed a face that bears an uncanny
   resemblance to Jesus Christ. All your friends and family
   see the pie and tell you that you've baked a miracle. You:

   A) Eat the pie.
   B) Take pictures of it on your iPhone and post it all over
       social media.
   C) Offer it up for auction on eBay, starting price $10,000.
   D) Call the Vatican, allow the pie to congeal, and then
       place it in a well-protected glass display case.

7. You wake up from having had an intense nightmare that
   someone was in your bedroom choking you. You can still
   feel the hands against your throat. Now that you think
   about it, when you were falling asleep you had a strange
   feeling that someone was in your room. You:

   A) Write a book about your psychic experience and self-
       publish it.
   B) Look around for evidence of a prowler and call the
       police if you find something suspicious.
   C) Call the evening news to warn the neighborhood
       about your home attack.
   D) Call the police right away.

8. One of your Facebook friends posts an article from a
   magazine you never heard of stating that eating too many
   vegetables, especially broccoli, can cause cancer. You:

   A) Empty your refrigerator bin of every vegetable and
       vow never to eat anything green again.

B) Repost the article on Facebook and your other social media accounts.

C) Dismiss the article as BS and stop following this friend's Facebook posts.

D) While having an anxiety attack thinking about all the vegetables you've consumed over your lifetime, you spend hours googling cases of people who died of vegetable cancer.

9. You are sitting in a bar having a good time with a half dozen friends. As the evening passes, one of your friends comments that the bar is empty and that you should head over to the one across the street because more people are there and she can hear the great music. Two friends agree and stand up to leave. The three remaining friends say, "No, thanks. I'm happy here." You equally enjoy the company of friends in both groups. You:

A) Leave the current bar. It's obviously unpopular and you may be labeled a loser if you stay there.

B) Before making a decision, you hold everyone up by going on your smartphone and comparing online reviews of the two bars.

C) Go home. The friends who are leaving for the other bar ruined everything.

D) Stay right where you are since you are having a good time and prefer being in a quiet place where you can hear what your friends are saying.

10. Al Gore becomes spokesperson for a scientific team from Harvard, which has drawn the conclusion that climate change is occurring 20 percent faster than previously believed. You:

A) Tell your Harvard grad friends to stop donating to the Harvard alumni organization.

B) Stop using all products associated with gasoline and electricity.

C) Read the research findings for yourself and draw your own conclusion.

D) Write a letter to Al Gore also congratulating him on having invented the Internet.

11. You are walking in a park late at night and see what you think may be a suspicious person about to attack you. You:

A) Use your iPhone light to get a better look, before sprinting off.

B) Use your iPhone to get good pictures of the person for posting on Instagram.

C) Panic and then hurl your iPhone at the figure.

D) Call your BFF on your iPhone and describe the person to her as the suspicious character draws closer.

12. You are the head of a powerful organization and are on a retreat at a lake with your staff. You do *not*:

A) Organize role-playing games.

B) Discuss the future of your organization and everyone's role in it.

C) Set up required sporting outdoor activities, such as water skiing, so that everyone can see that you've been working out.

D) Bring along your faithful Labrador retriever to participate in the activities.

13. You are attending your weekly religious service. During the sermon the leader of the group says, "People with red hair are led by the devil. They must vanquish all their evil, carnal thoughts which they cannot control." Most of the congregation, which consists of blondes and brunettes, chants "amen" in unison. You:

A) Dye your hair red and hit the bars.

B) Laugh out loud until your chest hurts.

C) Walk out with your redheaded friend, who has started to sob.

D) Dig up research studies proving that redheads are no more evil than people of any other hair color and send your findings to the leader.

14. Your boss, who has been leading a company meeting, writes on a whiteboard a conclusion and next steps to take. You think that her presumptions are faulty and that the next steps don't make sense. You are about to ask about it, but the boss announces this is officially the new company direction. Everyone around you enthusiastically nods and applauds. You:

A) Boo your boss and then toss your latte at her.

B) Applaud with everyone else.

C) Sit and sulk.

D) Applaud politely and then set up a private conversation with your boss for later in the day.

15. You attend the show of a famous mentalist performer whose act involves mind reading. He selects you from the audience to participate in his demonstration. After asking you a few bizarre questions, he is able to accurately state your name, age, occupation, and place of residence. The audience is enthralled. You:

A) Try to figure out how he discerned so much about you with so little detail.

B) Think the mentalist cheated and conferred with your mother about you before the show.

C) Believe the mentalist has superpowers.

D) Tell your friends who watched the show that the mentalist is a fake.

16. On TV, George Clooney, who has been your favorite actor for years, asks you to pledge money to help save the world's slug population, which he says is endangered. Without enough slugs, he expounds, our soil will deteriorate and thus devastate our food supply. You:

   A) Heckle the TV.
   B) Open up your checkbook right away and send $100 off to Save the Slugs.
   C) Run outside to collect every slug you can find and start breeding them.
   D) Change the channel and see if you can find a rerun of *ER*.

17. You're at the trendiest bar in LA. Many top actors, models, artists, writers, producers, musicians, and trendsetters are there. You somehow end up in the middle of a conversation with the likes of Mila Kunis, Scarlett Johansson, Demi Moore, Kate Upton, and Megan Fox. Demi says that what she has in her glass is the hottest new cocktail, the Whiskery Dingo. It costs $150 a shot but is totally worth it because two key ingredients in the mix — saliva from a dingo and the whisker of a desert mole rat — are really hard to come by in most LA bars. She swirls down the drink, smacks her lips, and makes a "yummy" sound. Mila, Scarlett, Kate, and Megan make a mad dash to the bar to get a Whiskery Dingo before the bar runs out of whiskers. You:

   A) Snap a pic of Demi with her glass still in hand and text it to all your friends, telling them that the Whiskery Dingo is the hottest drink ever.
   B) Wait it out and see which beauty "tosses her cookies" first.
   C) Join the women to plunk down $150 for a shot.

D) Get out your phone and order the premade Whiskery Dingo mix online, which sells for $300.

18. You've been standing in line at the Department of Motor Vehicles for an hour. The line hasn't moved an inch in fifteen minutes. You:

A) Say to the person behind you, "It feels like I've been in this line forever."
B) Text your mother, "I'd rather have a Whiskery Dingo than stay in this line another minute."
C) Bitch to the director of the DMV branch that her team's service "sucks."
D) Read a book on your Kindle.

19a. For men (women should go to the alternate question, 19b, below): Your wife, a hairdresser, tells you your newly grown beard is grotesque and is giving her rash burn, and that you need to shave it off right away. You look in the mirror. Part of you thinks she may be right; the other part is convinced she's only saying this because she loves to criticize you. You:

A) Go to the barber and get the beard professionally groomed.
B) Assert yourself, man — let the beard hang out as if you're a hipster in Brooklyn.
C) Ask the best-looking woman in your office for her opinion, and then decide.
D) Shave it off right away.

19b. For women (men should go to the alternate question, 19a, above): Lately you've been wondering if your teeth are yellowing and if it's worth getting them whitened. You ask your husband, a dentist, for his opinion. He looks into your mouth, thinks for a moment, and answers: "Your teeth look perfect, as always; I'd leave them alone." You:

A) Scream at him that he's a liar and is just saying that to make you feel good.
B) Get a second opinion from your friend Jerry, the dentist who happens to live next door.
C) Run out and purchase whitening strips from the local drugstore.
D) Thank your husband and get yourself a well-deserved latte for being so conscientious with your oral hygiene.

20. Picture being thrown via time warp back into the era of the Crusades. The pope makes a commanding speech, urging you to help reclaim the Holy Land. Everyone you know giddily signs up and heads off toward Jerusalem. The local clergy threaten to imprison you if you don't go right now. You:

A) Start singing "Onward Christian Soldiers" and join the troops.
B) Hold a bed-in and sing "Give Peace a Chance."
C) Try to get discharged by dressing up as a lunatic.
D) At the risk of getting caught and facing punishment, sneak off and run as far away as you can.

Here are the assessment answers showing the highest PI. Give yourself a point for each one you chose:

1. C
2. C
3. D
4. B
5. B
6. A
7. B

8. C
9. D
10. C
11. A
12. C
13. C
14. D
15. A
16. D
17. B
18. D
19. D (for both the male, 19a, and female, 19b, questions)
20. D

Scoring:

17–20 = High PI. Your PI is off the charts!

11–16 = Average PI. Your PI is pretty strong, but sometimes all it takes is one misjudgment to find yourself careening off the rails.

0–10 = Low PI. You should *not* order anything online or join a cult.

How You Think and Your Perceptual Intelligence

No matter how you scored on the assessment, one thing is certain: the quiz lacks the emotions that you will experience by *living* those situations. Recognizing reality in the heat of the moment isn't always easy, as we've learned throughout this book. *Life is the true test of your PI.* Everything is subject to interpretation. The exact same image can have an infinite number of meanings in the real world. When it comes to trusting our senses — seeing, feeling, hearing, tasting, smelling, and touching — there is no such thing as 100 percent objectivity, because we need to think about what we experienced, even if for just a few moments. Therefore, you can improve you Perceptual Intelligence by focusing more on your intuition, ability to think critically, individual thinking, and emotions. All four are useful elements in fostering high PI. In an effort to help you cultivate them, I've outlined general techniques to help you become more mindful in those areas. I've indicated the numbers from the assessment that apply to each type of thinking so you can home in on your areas of improvement.

Focusing on Intuition to Boost PI: Questions 3, 4, and 5

I included only three intuitive questions to solve, numbers 3, 4, and 5, which makes this one revealing; you may be in some trouble if you missed any of these. Intuitive thinking, which we covered in the last chapter, is the most difficult ability to recognize and interpret under "battle conditions." Other people may question your intuitive decision. Here I used the examples of the warning signs of steering clear of a plane crash and what to do in the case of a contaminated water scare. In both instances, your intuition is acting like a guardian angel and giving you invaluable hints that could prevent catastrophe. Heeding a queasy feeling about your tap water can prevent you from making a terrible mistake, such as ingesting lead or arsenic.

Of course, intuitive signals may also have positive repercussions, such as leading to a creative idea for an innovation, a painting, a song, or a story. In question 4, in which Tom Hanks is an insurance pitchman, I challenged you to glean an intuitive thought: If Tom Hanks — the most trusted star in Hollywood — sings the praises of a company or product, might it then be a safe bet that the company's stock would rise? It's a controversial and intuitive thought that is also risky, but that's the point: it just might pay off big. Of course, one could also look to critical thinking to avoid falling for a celebrity pitch — but the three other options I provided clearly reveal low PI.

Exploiting the thoughts that bubble up in your head at just the right instant can save your life, change your career or finances for the better, and even help you find a lasting relationship. The people with high PI regarding intuitive thinking seem always to "guess correctly." But that's not exactly what's happening here. When faced with a situation that doesn't present a logical choice, they go with their *first gut instinct*. When you are in those situations, try it yourself. Chances are, you would not do any worse than you would with all the useless deliberation that went on before you made the choice, and you'll have a better chance at making the best decision.

Focusing on Critical Thinking to Boost PI:
Questions 1, 6, 7, 12, 15, and 16

In these questions, you were tested with regard to your ability to think critically when your mind and senses are telling you that suspect pieces of information are real. Below are the high-PI answers, with explanations:

If an attractive person accosts you at Starbucks with pamphlets, you'll want to shoo her or him away, despite any attraction or curiosity. Note that this situation can be very real: cults tend to use attractive people to lure others in.

If you were to see Jesus Christ in your pot pie, eat it before it gets cold.

If you wake up from a dream that someone strangled you, search for evidence of an intruder to ease your fears. If you don't find anything, chalk it up to a dark lucid dream. If you do find evidence, contact the police right away.

If you are a business leader running a retreat, the worst thing you could do is force your employees to do outdoor stuff they'll probably hate just so they can admire your physique and massage your ego.

If a mentalist does an amazing trick to reveal things about you, it can be fun to try to figure out how he did it. You probably won't, but it's far better than going to ridiculous extremes like calling him a crackpot or believing he has superpowers.

If George Clooney wants to save the slugs, he can go right

ahead — but don't you do anything about it. If he's an actor you happen to like, just watch *ER* reruns or his new films instead.

In the instances above, I challenged you to use your ability to think critically to decide how to act in these situations. If you have high PI, you will question what your brain is trying to convince you of. Critical thinking means taking a pause and distancing yourself from these fantasies to try to get to the truth. If reality can't be attained, engaging in critical thinking to achieve high PI means not being willing to sacrifice your credibility and reputation for something absurd and damaging. This can be difficult because the emotions involved (such as fear when it comes to a dream about being attacked) feel powerful and real. If you find you are overly susceptible in this area, you may wish to seek coaching, counseling, or other therapeutic assistance.

Focusing on Individual Thinking to Boost PI:
Questions 8, 9, 13, 14, 17, and 20

I designed these questions to test your resistance to pressure from others in a variety of situations. Do you go with the crowd, follow the soccer ball down the field with everyone else, or get sucked into believing a foolish notion because most people say it's true? Individual thinking means being able to separate yourself from the crowd when the facts just aren't there to support what's being said. Sometimes intuitive thinking and critical thinking play a part in individual thinking, but occasionally with misleading results. In the case of the former, a group of people might make it seem as if their intuitive thoughts (such as tapping into the supernatural) warrant your joining their cult. In the latter, if a group posts nonfactual propaganda, backed up by a ton of so-called facts, and ten thousand people click the "Like" button, you may be swayed into following the herd and believing fake news.

In question 8 I came up with the ridiculous notion of someone posting an article stating that vegetables cause cancer. The reason I did this was simple. This exaggerated example is not so far removed from some of the crazy things I've seen posted and reposted on social media pages, meaning that somewhere out there are people gullible (or neurotic) enough to believe pretty much anything. The answer is not so clear-cut, however, and I'll bet some of you who answered honestly chose D ("having an anxiety attack" and "googling cases of people who died of vegetable cancer"). The correct answer, C ("dismiss the article as BS and stop following this friend's Facebook posts"), reveals high PI because it demonstrates that you won't consider preposterous claims and won't waste your precious time looking at future articles from someone who has lost credibility. (Snopes.com can be a credible source for verifying Internet rumors.)

The bar question, number 9, tests how much your PI is manipulated by the pressure of going with the popular choice (i.e., the more crowded bar). The correct answer is D because there is no reason for you to leave the bar if you are already comfortable, if you like the friends who are remaining, and if you wish to hear what they have to say without the blaring music getting in the way. Answer B (going online to compare reviews) might have had value before heading out that evening, but in any case doing so now is not the right timing because you would be holding people up with your deliberation, thereby potentially creating unnecessary indecision and drama. (I know people who do this, and it's endlessly frustrating.)

In question 14, I sought to expose the nonindividual thinking that occurs in the workplace. Just because the boss writes something on the board doesn't mean it's true. (As evidence, just watch any random episode of *The Office* — an X-ray satire of work life in America and Britain.) The worst things to do are to let your boss go, follow the crowd, or embarrass yourself or your boss. Your best bet is to retain your latte, be professional, and confer with your boss in private.

For question 15, on a personal note, if one of my friends were in a bar with Mila, Scarlett, Demi, Kate, and Megan, I admit he would probably succumb and have whatever they were having (dingo saliva or not). The second answer, B, demonstrates high PI because by "waiting it out" you would be thinking individually (as well as thinking critically) by resisting group pressure.

In the last question about battling in the Crusades, number 20, I was testing your ability to stand out from the crowd and think individually while being out of time and facing pressure that might involve imprisonment, social banishment, or even physical harm. While it might be tempting to see whether the pope would give out a Section Eight (military discharge), the correct answer is *fleeing* — get the hell out of there as soon as possible to save yourself rather than fighting in a war without substance.

Focusing on Your Emotions to Boost PI: Questions 2, 10, 11, 18, and 19

Last, when faced with emotional situations, we sometimes go off the rails, and logic and reason are thrown out the window. In his benchmark work *Emotional Intelligence*, Daniel Goleman refers to this as the "cave brain"; our emotions take over in high-pressure situations, and we revert to primitive thinking and reactions.

Question 2: Sports fans have been known to regress when their teams win or lose; they overreact to things that are beyond their control and that don't affect their daily lives. If you are an adult sleeping in a team football helmet in the middle of June, you may need some separation from ESPN — if not some coaching or therapy.

Question 10: Social and political issues are rife with emotion. Whether you are a liberal or a conservative, you have a stake in what happens to our planet. It doesn't matter that Al Gore, a Democrat, is the bearer of the cataclysmic news: if scientific fact from a trusted source (such as Harvard) indicates doom and gloom about our environment, you should pay attention.

Question 11: Many people consider their phone their most prized possession, so don't hurl it anywhere. Instead, use it as a tool when you need it most: shine a light and help discern fact from fiction, even when you are filled with emotion (such as terror).

Question 18: It's easy to go bonkers when faced with bureaucracy and when you feel like your time is being wasted. As time passes in these situations, we all start to get hot under the collar. The best bet, rather than assigning blame to the often sloth-like, job-protected civil servants working at the DMV, is to distract yourself by reading. Or to do a puzzle. Or to doodle funny caricatures of the people waiting with you. Do whatever floats your boat and helps pass the time. In this situation it would be prudent to use electronics to speed up your perception of time lapsing.

Question 19a (for men): I provided a couple of clues here as to why you should shave off the beard rather than attempt to groom it and "prove you are right." The first is that, although your wife could have a bias because she doesn't like beards in general or because as your wife she has some preconceived notions about you, she is a hairdresser in this scenario, indicating that she probably knows something about physical presentation. Second, in the question I intentionally reference that part of you has doubts about the beard, which means you are probably reacting emotionally to your wife's criticism.

Question 19b (for women): Self-perceptions aren't always the most accurate perceptions. A husband might be tempted to say what he thinks his wife wants to hear, but in this case the husband did make the effort of looking into her mouth before answering. He's also a dentist. The correct answer — the unemotional one — is that the wife should thank her husband for the compliment and move on. Going to Jerry (the dentist neighbor next door) for a second opinion (option B) would be downright insulting, and purchasing whitening strips (option C) shows a lack of restraint. If you are so desperate to whiten your teeth, despite your dentist husband's assessment, at least have the courtesy of telling him you disagree and asking him for his professional recommendation (which, in this scenario, is probably more effective than buying the product in the store).

Curbing emotions in stressful situations is impossible for some people, and I don't suggest for a second that a shortcut exists to help you if this is a chronic issue. However, when your cave brain gets the better of you and your Perceptual Intelligence bucket is empty, *resist reacting at all costs*. At a minimum, take three deep breaths or even a day to sleep on it. You'll thank me later.

Epilogue

PI: Your Final Perception

Now that you've had a chance to digest sixteen chapters on Perceptual Intelligence and taken the assessment — which, by the way, nearly qualifies you as an expert — what do you plan to do with all this knowledge? Will you think twice before purchasing a product tweeted about by a celebrity? Will you write down the amazing idea for an invention that came to you out of thin air while you were running on the treadmill or elliptical — and maybe even see if it's possible to execute it? If you find out about a Reuben sandwich with the pope's likeness burned onto it, do you promise me you will resist bidding $20,000 for it?

Skepticism, logic, and emotional intelligence are your greatest allies in detecting and correctly interpreting the truth about what you experience, especially when the signals are mixed, the world is muddled, and your senses seem to be failing.

On the other hand, intuition — your sixth sense — is equally valuable, if you tune in to the right wavelengths. Will you win the lottery if you dream about certain numbers and try to act on this "intuition"? The odds are against it. But if a tingle shoots through you for no logical reason, there is no harm in *exploring* it and then assessing the risk versus the value in testing out that gut feeling. Who knows? Maybe it will lead you down a new career path or give you the spark to finally write your novel. Just don't wait too long to make your decision: we know how fleeting time can be, as we established in chapter 14. And kudos to you for tuning out your smartphone long enough to read through this entire book, since we know from that same chapter that reliance on electronics messes with our sense of time.

With that, until next time, I welcome hearing your thoughts about the concept of Perceptual Intelligence and your personal experiences with it, as well as requests for anything else you'd like to hear about from me in the future. You may contact me through my website at www.PerceptualIntelligence.com. Information will also be available on the website about my speaking tour.

In the meantime, I offer you this closing contrarian thought on Perceptual Intelligence, and why discerning reality from fantasy isn't 100 percent necessary all the time:

Your eyes are your witness to the real world — but without imagination you could not do justice describing what you see.

P.S.: To be notified about my Perceptual Intelligence tour dates and cities, text the word *TOUR* to 310-598-2885.

Acknowledgments

As a medical student during my ob-gyn rotation, I delivered twenty babies. With this book it's now twenty-one. This healthy bundle of joy you hold in your hands wouldn't have been possible without my team assisting me in the delivery room.

In order of shoe size...

My agent, Gordon Warnock, who believed in me and my idea for this book. You were a superhero who swooped down from the sky to save the project from drowning. Gary M. Krebs, my partner in this crime, who understands how I think and, most importantly, my humor — it was fun working with you! David Nayor, for seeing the tree in the seed. Jonathan Franks, thank you for your catalytic efforts and unshakable belief in the book when it was an idea. Montel Williams, for being behind me from the start — you are an incredible inspiration. Harvey-Jane Kowal, who is always listening and advising from her tower of experience. Karen Kosztolnyik, for covertly helping me while you were behind enemy lines — mission accomplished! Pam Shriver, for opening up and sharing your insights. Dr. Marvin Galper, for conveying your expertise and harrowing experiences. Matt Torrington, MD, for your insights about dopamine and addiction. Thank you, Georgia Hughes at New World Library, for being completely behind this book and turning the perception into reality. I sincerely appreciate Monique Muhlenkamp, Munro Magruder, Kristen Cashman, and the rest of the New World Library team for throwing themselves behind this book to ensure it reaches a large audience of hungry readers. Do I thank Mimi Kusch for her X-Acto knife copyediting handiwork that left many severed manuscript bits writhing on the floor? Of course! Thank you, Mimi! We owe a great debt to Richard Fox, who pointed us in the right direction with art research. Thank you, Blackstone Audio, who made me sound really good in the audiobook of *Perceptual Intelligence* (and was tons of fun for me to narrate with lots of "personality" for listeners).

A big shout-out goes to fellow Dartmouth Medical School alum and friend John Kennedy, MD, for lending his ear and advice based on his experience as a published author. A mega-thank-you to prolific author and fan/encourager Boze Hadleigh, for always seeing that I could do this. Thank you, Ana Zamalloa, for sharing Jerry's story in Peru. Thank you, Carol Gross of the United States Olympic Committee, for your assistance. Ari Galper, I am so grateful for meeting you at camp when we were pimply twelve-year-olds and I so value your friendship and sage advice over the decades — thank you, 2ARI! Dan Kennedy, thank you for your invaluable writing and strategic help over many years. A big-block thank-you to Susan and Norm Nelson, Geoff Stunkard, Bob McClurg, and Leon Perahia for confirming my gut feeling that Plymouth's muscle car color, Curious Yellow, was named after a Swedish X-rated film.

To my wife, Selina, who has been at my side through everything, there are no words to express my gratitude, but I know, after twenty-four years of marriage, that there are plenty of flowers available for thank-yous. Thank you to my twin daughters, who, at the many Friday-night-dinner recounting of the week's ups and downs during our "roses and thorns" discussions, listened with interest and asked questions about the soap opera that is book publishing.

Thank you to my caring, expert, and genuinely patient and dedicated staff — my second family — at my private practice in Beverly Hills, many of whom have been with me for well over a decade.

A special note of thanks to longtime friend Ron Thomson, who assisted with Shakespeare before the clock struck midnight. Don't go changing!

Last, I'd like to thank all those city workers who fix the potholes in our roads. It has nothing to do with the book, but no one ever thanks them. So here goes: "I thank you, the dedicated women and men in orange who keep our roads and radials safe. On behalf of millions of motorists whose coffee never got jarred and spilled all over them on the way to work because of your commitment to smooth asphalt, I thank you all."

About the Author

An expert in human perception and one of the world's leading authorities on Keratoconus, laser vision correction (LASIK), and dry eye treatments, Brian Boxer Wachler, MD (a.k.a. Dr. Brian), has devoted his career to the field of eye surgery. For two decades he has been a pioneering doctor with a career in clinical, academic, and research settings. His vast area of expertise extends into understanding how people think and how the mind works.

He is currently the director of the eponymous Boxer Wachler Vision Institute in Beverly Hills and a staff physician at Los Angeles's famed Cedars-Sinai Medical Center.

Dr. Brian has made countless contributions to the field of ophthalmology and ophthalmic surgery and received thirty-nine honors and awards over his career. He is credited with transforming the treatment of Keratoconus through his breakthroughs in noninvasive Holcomb C3-R® crosslinking and Intacs® inserts; he is known inside and outside medical circles as the "Keratoconus Guru" and is the author of three books on the subject. In 2010 Dr. Brian was recognized with the Jules Stein Living Tribute Award for inventing the Holcomb C3-R® treatment for Keratoconus, alongside US bobsled driver Steven Holcomb (the procedure's namesake) for winning Olympic gold after having the procedure. This procedure has saved thousands of Keratoconus sufferers from needing to undergo invasive and painful cornea transplants. More details about Keratoconus, Steven Holcomb's story, and Holcomb C3-R® may be found at www.KeratoconusInserts.com. You can also read Steven's book, *But Now I See: My Journey from Blindness to Olympic Gold* (BenBella Books, 2012).

Olympic gold medal winner Steven Holcomb and Dr. Brian

Dr. Brian is a recognized innovative leader in the treatment of glasses and contact lens–related refractive errors — vision problems that happen when the shape of the eyes prevents them from focusing properly — such as myopia (nearsightedness), hyperopia (farsightedness), astigmatism, and presbyopia (the need for reading glasses). Dr. Brian wrote one of the most popular reports on LASIK, the most widely performed type of refractive surgery. Dr. Brian's work on LASIK led to the formation of industry-wide guidelines to help make LASIK one of the safest procedures in medicine. Dr. Brian pioneered a procedure to treat brown spots, or freckles, on the whites of eyes as well as bloodshot, or red, eyes. His procedure also treats pterygium, pinguecula, and nevus of Ota. He also has extensive experience in advanced cataract surgery. More information about all his procedures is at www.BoxerWachler.com.

Dr. Brian has written eighty-four medical articles and twenty book chapters, and has delivered 276 scientific presentations. He has written four other books and is an inventor, which includes two approved patents for dry eye treatments. He has participated in fifteen FDA clinical trials evaluating new technologies. A board-certified ophthalmologist, he is a member of the American Academy of Ophthalmology. Dr. Brian holds leadership positions in numerous organizations, including being a medical editor for WebMD. He has been featured on all the major TV networks and has been written about in numerous newspapers and magazines. For more about his media coverage, please visit his IMDB page.

Dr. Brian lives in Los Angeles with his wife, to whom he has been married since 1993. Of note, he is one of a select number of men in the country with a maiden name; when they married, he and his wife combined each other's last names to create Boxer Wachler. They have loving preteen twin daughters who are a delight. Dr. Brian still competitively rows (which he did in college) and often employs his college debate team skills with his daughters over limiting their iPad time.

To be notified about Dr. Brian's Perceptual Intelligence speaking tour dates and cities, text the word *TOUR* to 310-598-2885.

Notes

Introduction

Page 1 *In 2009 thousands of visitors flocked:* "Virgin Mary Seen in Tree Stump in Limerick," *Belfast Telegraph*, September 7, 2009.

Page 1 *It has been approximated, from various surveys:* C. Taylor, "How Many People Actually Manage to Have an Out of Body Experience?," June 14, 2014, Out-of-body-experience.info/how-many-people-had-an-obe.

Page 1 *According to reports from the National UFO Reporting Center:* National UFO Reporting Center, data from 2016, www.nuforc.org/webreports /ndxevent.html.

Page 3 *Anyone living with Charles Bonnet Syndrome:* Maureen A. Duffy, "Charles Bonnet Syndrome: Why Am I Having These Visual Hallucinations?," *Vision Aware*, www.visionaware.org/info/your-eye-condition /guide-to-eye-conditions/charles-bonnet-syndrome/125.

Page 4 *A medical condition known as synesthesia:* Siri Carpenter, "Everyday Fantasia: The World of Synesthesia," *The American Psychological Association Journal* 32, no. 3 (March 2001), www.apa.org/monitor/mar01 /synesthesia.aspx.

Page 5 *insects can have up to twenty-five thousand:* "Evolution of the Insect Eye," University of Minnesota Duluth newsletter, www.d.umn.edu/~olseo1/0 /Evolution/insects.html.

Page 5 *Chickens, on the other hand, don't see:* Thomas J. Lisney et al., "Behavioural Assessment of Flicker Fusion Frequency in Chicken *Gallus gallus domesticus*," *Science Direct* 51, no. 12 (June 2011), www.sciencedirect .com/science/article/pii/S0042698911001519.

Page 6 *In their excellent book:* L. Michael Hall and Bob G. Bodenhamer, *The User's Manual for the Brain*, vol. 2 (New York: Crown, 2003).

Chapter 1. Perception's Seat

Page 10 *In 1999 Andy and Larry Wachowski created:* Evan Katz, "Am I Dreaming? The Matrix and Perceptions of Consciousness," January 18, 2013, Philfilm rhodes.blogspot.com/2013/01/am-i-dreaming-matrix-and-perceptions -of_18.html.

Page 11 *The film's executive producer Andrew Mason:* Paul Martin, "Interview with Andrew Mason," *MatrixFans.net*, February 13, 2012, www.matrixfans .net/interview-with-andrew-mason-executive-producer-from-the-matrix -1999/#sthash.A7gmGxph.dpbs.

Page 11 *Neuroscientist Sam Harris describes it as:* Sam Harris, "The Self Is an Illusion," YouTube video, September 16, 2014, www.youtube.com /watch?v=fajfkO_Xolo.

Page 12 *The Merriam-Webster online dictionary defines: Merriam-Webster's Learner's Dictionary,* s.v. "Perception," accessed May 18, 2017, www.learners dictionary.com/definition/perception.

Page 13 *In his book* An Inquiry into the Human Mind*:* Thomas Reid, *An Inquiry into the Human Mind: On the Principles of Common Sense* (Charleston, SC: Nabu, 2010), reproduction of the original text.

Page 14 *Philosophers continue to debate this:* Jim Baggott, "Quantum Theory: If a Tree Falls in the Forest...," *OUPblog*, Oxford University Press blog, February 14, 2011, Blog.oup.com/2011/02/quantum.

Page 15 *Our sensory organs contain receptor cells:* "Receptors," *CNS Clinic*, www .humanneurophysiology.com/receptors.htm.

Page 15 *Then the already modified information:* Sally Robertson, "What Does the Thalamus Do?," *News Medical Life Sciences*, www.news-medical.net /health/What-does-the-Thalamus-do.aspx.

Page 16 *Let us not forget the cerebral neocortex:* "Neocortex (Brain)," *Science Daily*, www.sciencedaily.com/terms/neocortex.htm.

Page 16 *Reality is real, but what we see, hear, feel, touch:* Bahar Gholipour, "Made Up Purely by the Brain," *Brain Decoder*, Braindecoder.com/post/up-to-90 -of-your-perception-could-be-made-up-purely-by-the-brain-1104633927.

Page 18 *At Ohio State University, scientists have used:* Emily Caldwell, "Scientist: Most Complete Human Brain Model to Date Is a 'Game Changer,'" Ohio State University news release, August 18, 2015, news.osu.edu/news /2015/08/18/human-brain-model.

Page 18 *All this comes on the heels of:* Emily Underwood, "More Than $100 Million in New BRAIN Funds," *Science*, October 2, 2015, www.sciencemag.org/news/2015/10/more-100-million-new-brain-funds.

Page 18 *Yet, as the* New York Times *reported in 2014:* James Gorman, "Learning
     How Little We Know about the Brain, *New York Times,* November 10, 2014,
     www.nytimes.com/2014/11/11/science/learning-how-little-we-know-about
     -the-brain.html.
Page 19 *Although our brains are constantly interpreting inputs:* Rebecca Tan,
     "9 Unanswered Questions about the Human Brain," *South China Morning
     Post,* May 9, 2016, www.scmp.com/lifestyle/health-beauty
     /article/1941658/9-unanswered-questions-about-human-brain.

Chapter 2. Mind Over (and Under) Matter

Page 21 *According to a 2008 survey:* Salynn Boyles, "86 Billion Spent on Back,
     Neck Pain," *WebMD,* February 12, 2008, www.webmd.com/back-pain
     /news/20080212/86-billion-spent-on-back-neck-pain.
Page 21 *a particularly challenging disease to treat:* National Multiple Sclerosis
     Fact Sheet, www.nationalmssociety.org/NationalMSSociety/media
     /MSNationalFiles/Brochures/Brochure-Just-the-Facts.pdf.
Page 22 *by no means the only celebrity to have been treated:* Brian Krans,
     "Famous Faces of MS," *Healthline,* February 8, 2017, www.healthline.com
     /health-slideshow/famous-people-with-ms.
Page 22 *You may have heard the story of:* Lisa Stein, "Living with Cancer: Kris
     Carr's Story," *Scientific American,* July 16, 2008, www.scientificamerican
     .com/article/living-with-cancer-kris-carr.
Page 23 *By being crazy sexy in her approach:* Kris Carr, *Crazy Sexy Cancer Tips*
     (Guilford, CT: Skirt!, 2007).
Page 23 *According to a recent Johns Hopkins study:* "Don't Worry, Be Healthy,"
     Johns Hopkins Medicine, press release, July 9, 2013, www.hopkinsmedicine
     .org/news/media/releases/dont_worry_be_healthy.
Page 23 *Substantial evidence reveals:* Jessica Carretani, "The Contagion of
     Happiness," *Harvard Medicine,* hms.harvard.edu/news/harvard-medicine
     /contagion-happiness.
Page 24 *In one famous study by Jon Kabat-Zinn:* "Jon Kabat-Zinn, PhD," *The
     Connection,* theconnection.tv/jon-kabat-zinn-ph-d.
Page 24 *Dr. Daniel Siegel, author of the groundbreaking book:* Daniel J. Siegel,
     Mindsight: The New Science of Personal Transformation (New York:
     Bantam, 2010).
Page 27 *Some scientists attribute this phantom pain:* Susha Cheriyedath, "What
     Is a Phantom Limb?," www.news-medical.net/health/What-is-a-Phantom
     -Limb.aspx.
Page 27 *In order to be officially diagnosed with* hypochondriasis: "Hypochondri-
     asis," *Cleveland Clinic,* my.clevelandclinic.org/health/articles/hypo
     chondriasis.

Page 28 *If you're a hypochondriac, you can feel somewhat better:* Brian Dillon, "The Pain of Fame," *Wall Street Journal,* January 16, 2010, www.wsj.com /articles/SB10001424052748704281204575003570232360564.

Page 28 *Some experts now theorize that the barrage:* Chris Weller, "Dr. Google Breeds Hypochondria by Scaring People into Thinking the Worst," *Medical Daily,* May 7, 2015, www.medicaldaily.com/dr-google-breeds -hypochondria-scaring-people-thinking-worst-332316.

Page 29 *A rash of recent viral posts inaccurately said:* "CDC Recommends Mothers Stop Breastfeeding to Boost Vaccine Efficacy?," Snopes.com, January 21, 2015, www.snopes.com/medical/disease/cdcbreastfeeding.asp.

Chapter 3. What You See Is Not What You Get

Page 31 *Separate from magic exists:* "The Top 10 Mentalists," *Mentalist Central,* January 9, 2015, www.mentalismcentral.com/top-10-mentalist.

Page 32 *Other individuals are reported to have:* Ingo Swann website, www .ingoswann.com.

Page 32 *In the late 1970s the Defense Intelligence Agency:* Central Intelligence Agency website, https://www.cia.gov/library/readingroom/document /cia-rdp96-00787r000200130005-3.

Page 32 *At the time the government believed:* Central Intelligence Agency website, https://www.cia.gov/library/readingroom/document/cia -rdp79-00999a000300030027-0.

Page 32 *Then, of course, there are the skeptics:* Adam Higginbotham, "The Unbelievable Skepticism of the Amazing Randi," *New York Times Magazine,* November 7, 2014, www.nytimes.com/2014/11/09/magazine/the -unbelievable-skepticism-of-the-amazing-randi.html.

Page 32 *Margaret Thatcher once said of her detractors:* Dean Gualco, *The Great People of Our Time* (Bloomington, IN: iUniverse, 2008).

Page 34 *To the question of whether aliens exist:* Carl Sagan, *The Demon-Haunted World: Science as a Candle in the Dark* (New York: Random House, 1997).

Page 34 *People with symptoms of sleep paralysis:* "Sleep Paralysis," *WebMD,* www.webmd.com/sleep-disorders/guide/sleep-paralysis#1.

Page 34 *This is when their "dark lucid dreams":* Rebecca Turner, "Are Alien Abductions Real — or Dark Lucid Dreams?," www.world-of-lucid-dreaming .com/are-alien-abductions-real.html.

Page 34 *the extreme version of "lucid dreams":* Jake Rossen, "The Dark Side of Lucid Dreaming," Van Winkle's, September 25, 2016, https://vanwinkles .com/lucid-dreamings-dark-side.

Page 37 *Pablo Picasso once said:* Pablo Picasso, "Statement to Marius De Zayas,"

1923, www.learn.columbia.edu/monographs/picmon/pdf/art_hum
_reading_49.pdf.

Page 38 *For centuries, artists have used specific colors:* Esther Inglis-Arkell, "Why
Certain Color Combinations Drive Your Eyeballs Crazy," *Gizmodo*, January 13, 2013, io9.gizmodo.com/5974960/why-certain-color-combinations
-drive-your-eyeballs-crazy.

Page 38 *When it comes to art, our brains:* Chuck Close website, chuckclose.com.

Page 39 *Lest we think that painters are the only ones:* Stephen Sondheim and
James Lapine, *Sunday in the Park with George*, book version (Applause
Theatre & Cinema Books, 2000).

Page 40 *One of the paintings most etched in our minds:* Arthur Lubow, "Edvard
Munch: Beyond the Scream," *Smithsonian*, March 2006, www.smithsonian
mag.com/arts-culture/edvard-munch-beyond-the-scream-111810150.

Page 40 *Another Expressionist painter, Vincent van Gogh:* "Van Gogh's Mental
and Physical Health," *Van Gogh Gallery*, www.vangoghgallery.com/misc
/mental.html.

Page 41 *In his work* Phaedo: Plato, *Phaedo* (New York: Oxford University Press,
2009).

Page 42 *In one study, a group of people:* Karen Brakke, "Ponzo," *Online Psychological Laboratory*, opl.apa.org/Experiments/About/AboutPonzo.aspx.

Page 43 *Can you imagine trying to drive:* Marc Abrahams, "Experiments Show
We Quickly Adjust to Seeing Everything Upside-Down," *Guardian*, November 12, 2012, www.theguardian.com/education/2012/nov/12/improbable
-research-seeing-upside-down.

Page 43 *At the conclusion of a speech:* Amanda Enayati, "The Power of Perceptions: Imagining the Reality You Want," *CNN*, April 14, 2012, www.cnn
.com/2012/04/11/health/enayati-power-perceptions-imagination.

Chapter 4. Out of Body or Under the Ground

Page 45 *"I don't believe in an afterlife":* Woody Allen, "Conversations with
Helmholtz," *Getting Even* (New York: Vintage, 1978).

Page 45 *Among those medical practitioners who believe:* Raymond Moody Jr., *Life
after Life: The Bestselling Original Investigation That Revealed "Near-Death
Experiences"* (San Francisco: HarperOne, 2015).

Page 45 *Dr. Mario Beauregard, coauthor:* Mario Beauregard and Denyse
O'Leary, *The Spiritual Brain: A Neuroscientist's Case for the Existence of the
Soul* (San Francisco: HarperOne, 2008).

Page 45 *Dr. Mary Neal, author:* Mary Neal, *To Heaven and Back: A Doctor's
Extraordinary Account of Her Death, Heaven, Angels, and Life Again*
(Colorado Springs, CO: WaterBrook, 2012).

Page 45 *Dr. Jeffrey Long, author of:* Jeffrey Long, with Paul Perry, *Evidence of the Afterlife: The Science of Near-Death Experiences* (San Francisco: HarperOne, 2011).

Page 46 *an organization that has collected:* Near Death Experience Research Foundation (NDERF) website, www.nderf.org.

Page 46 *To find statistics on NDEs:* Tara MacIsaac, "How Common Are Near-Death Experiences?: NDEs by the Numbers," *Epoch Times*, June 23, 2014, www.theepochtimes.com/n3/757401-how-common-are-near-death-experiences-ndes-by-the-numbers.

Page 46 *Sam Harris has speculated about the idea of consciousness:* See Sam Harris, *Waking Up: A Guide to Sprituality Without Religion* (New York: Simon & Schuster, 2014).

Page 47 *There is the story of four-year-old:* Todd Burpo and Lynn Vincent, *Heaven Is for Real: A Little Boy's Astounding Story of His Trip to Heaven and Back* (Nashville, TN: Thomas Nelson, 2010).

Page 47 *More startling is the tale told:* Eben Alexander, *Proof of Heaven: A Neurosurgeon's Journey into the Afterlife* (New York: Simon & Schuster, 2012).

Page 48 *When a patient stops breathing:* "Special Report: When Is Your Patient Dead?," *Medscape,* www.medscape.com/viewcollection/32925.

Page 49 *In 2013 the University of Michigan:* Rob Stein, "Brains of Dying Rats Yield Clues about Near-Death Experiences," on *All Things Considered,* NPR, August 12, 2013, www.npr.org/sections/health-shots/2013/08/12/211324316/brains-of-dying-rats-yield-clues-about-near-death-experiences.

Page 49 *In a different study of rats:* Shantell Kirkendoll, "Study: Near-Death Brain Signaling Accelerates Demise of Heart," *University Record,* University of Michigan newsletter, April 10, 2015, record.umich.edu/articles.

Page 50 *There are other potential explanations:* Francis Grace, "The Science of Near-Death Experiences," *CBSNews,* April 18, 2006, www.cbsnews.com/news/the-science-of-near-death-experiences.

Page 50 *The late Elisabeth Kübler-Ross, psychiatrist:* Elisabeth Kübler-Ross website, www.ekrfoundation.org.

Page 53 *Celebrities aren't immune to viruses:* Near Death Experiences of the Hollywood Rich and Famous, website, www.near-death.com/experiences/rich-and-famous.html.

Page 54 *I saw, like, dark masks crushing:* Bill Clinton, "How Clinton Recovered from Surgery," *ABC News,* October 28, 2004, abcnews.go.com/Primetime/clinton-recovered-surgery/story?id=207370.

Page 55 *In one famous case, a Dutch patient's:* G. M. Woerlee, "The Denture Man NDE," *Near Death Experiences,* www.neardth.com/denture-man.php#lommel.

Page 56 *In early 2014 researchers at the University of Ottawa:* Andra M. Smith and Claude Messier, "Voluntary Out-of-Body Experience: An fMRI Study," *Frontiers in Human Neuroscience,* February 10, 2014, journal.frontiersin .org/article/10.3389/fnhum.2014.00070/full.

Page 56 *According to the* Psychology Dictionary: *Psychology Dictionary,* s.v. "What Is Kinesthetic Imagery?," psychologydictionary.org/kinesthetic -imagery.

Page 57 *This phenomenon, sometimes referred to as "blindsight":* David Robson, "Blindsight: The Strangest Form of Consciousness." *BBC Future,* September 28, 2015, www.bbc.com/future/story/20150925-blindsight-the-strangest -form-of-consciousness.

Page 57 *Dr. Ken Paller, a professor of psychology:* Ken Paller and Satoru Suzuki, "Consciousness," report, Northwestern University, faculty.wcas.north western.edu/~paller/Consciousness.pdf.

Page 57 *A man known to the medical world as "Patient TN":* Graham P. Collins, "Blindsight: Seeing without Knowing It," *Scientific American,* blog, April 22, 2010, blogs.scientificamerican.com/observations/blindsight-seeing -without-knowing-it.

Chapter 5. Vanity Games

Page 60 *The Sochi Olympics were Putin's Games:* Neil Tweedie, "The Dark Side of Vladimir Putin's Winter Olympic Games," *Telegraph,* February 1, 2014, www.telegraph.co.uk/sport/othersports/winter-olympics/10610000 /The-dark-side-of-Vladimir-Putins-Winter-Olympic-Games.html.

Page 61 *In the post-Communist age:* Ilan Berman, "Putin's Olympic Corruption," *USA Today,* February 20, 2014, www.usatoday.com/story/opinion /2014/02/20/putin-olympics-sochi-corruption-russia-column/5655815.

Page 61 *and his image isn't just "shtick":* "Vladimir Putin's Tough Guy Act Is Just 'Shtick' Says Barack Obama," *Telegraph,* February 7, 2014, www.telegraph .co.uk/news/worldnews/barackobama/10623452/Vladimir-Putins -tough-guy-act-just-a-shtick-says-Barack-Obama.html.

Page 62 *which even included the prosecution of the band:* "Russia: Punk Band Arrested after Protesting Putin," *Freemuse,* Freemuse.org/archives/1914.

Page 62 *Putin knew this tactic would strike a chord:* Rebecca Perring, "Vladimir Putin 'Wants' to Reinstate Russia's Royal Family and Bring Back the Tsars," *Express,* June 24, 2015, www.express.co.uk/news/world/586470/Russia -royal-family-Vladimir-Putin-reinstate-Tsar-Nicholas-Second-Romanov.

Page 63 *Putin and Trump met face-to-face for the first time*: Emily Shugerman, "Putin Points at Journalists and Asks Trump 'Are These the Ones Hurting You?' during Press Conference," *Independent,* July 7, 2017,

www.independent.co.uk/news/world-0/us-politics/trump-putin-press
-journalists-meeting-russia-president-points-which-ones-insulting
-you-a7830046.html.

Page 63 *When Russia first bid for the 2014 Games:* Will Stewart, "A $51 Billion
'Ghetto': Extraordinary Images Show Vladimir Putin's Sochi Olympic
Park Lying Desolate and Abandoned One Year after Most Expensive
Games in History," *Daily Mail,* February 6, 2015, www.dailymail.co.uk
/news/article-2941216/Extraordinary-images-Vladimir-Putin-s-Sochi
-Olympic-park-lying-desolate-abandoned.html.

Page 64 *It's now widely known that Russian officials:* Joyce Chen, "The Russian
Olympic Doping Scandal Explained: 5 Things to Know," *Us,* May 13, 2016,
www.usmagazine.com/celebrity-news/news/the-russian-olympic-doping
-scandal-explained-5-things-to-know-w206469.

Page 65 *Never before has the term* political football: Jeremy Quittner, "Patriot's
Owner Robert Kraft Still Wants Putin to Give Back His Super Bowl Ring,"
Fortune, February 6, 2017, fortune.com/2017/02/06/patriots-owner
-kraft-putin-ring.

Page 66 *Poll after poll shows:* Ian H. Robertson, "The Danger That Lurks inside
Putin's Brain," *Psychology Today,* blog, May 17, 2014, www.psychology
today.com/blog/the-winner-effect/201403/the-danger-lurks-inside
-vladimir-putins-brain.

Page 66 *Putin's well-photographed macho endeavors:* "Vladimir Putin's Macho
Stunts," *The Economist,* May 26, 2015, www.economist.com/node/21652100.

Page 67 *In June 1992 Bill Clinton donned sunglasses:* David Zurawick, "Bill
Clinton's Sax Solo on 'Arsenio' Still Resonates Memorable Moments,"
Baltimore Sun, December 27, 1992. http://articles.baltimoresun.com/1992
-12-27/features/1992362178_1_clinton-arsenio-hall-hall-show.

Chapter 6. Let's Get Physical

Page 70 *Numerous studies have shown that:* "Physical Activity Reduces Stress,"
Anxiety and Depression Association of America, online forum, www.adaa
.org/understanding-anxiety/related-illnesses/other-related-conditions
/stress/physical-activity-reduces-st.

Page 70 *During exercise, the body unleashes:* Kristin Domonell, "Why Endorphins (and Exercise) Make You Happy," *CNN,* January 13, 2015, www.cnn
.com/2016/01/13/health/endorphins-exercise-cause-happiness.

Page 72 *You may have heard of gray matter:* Susan Scutti, "Brain Facts to Know
and Share: Men Have a Lower Percentage of Gray Matter Than Women,"
Medical Daily, July 10, 2014, www.medicaldaily.com/brain-facts-know
-and-share-men-have-lower-percentage-gray-matter-women-292530.

Page 72 *A recent study revealed that white matter:* Arthur F. Kramer, "Enhancing Brain and Cognitive Function of Older Adults through Fitness Training," *Journal of Molecular Neuroscience* 20, no. 3 (February 2003), www.researchgate.net/publication/9087834_Enhancing_Brain_and _Cognitive_Function_of_Older_Adults_Through_Fitness_Training.

Page 73 *In one experiment, twenty-three volunteers were asked:* Christol Koch, "Looks Can Deceive: Why Perception and Reality Don't Always Match Up," *Scientific American*, July 1, 2010, www.scientificamerican.com/article /looks-can-deceive.

Page 74 *In today's sports world there probably isn't:* AJ Adams, "Seeing Is Believing: The Power of Visualization," *Psychology Today*, December 3, 2009, www.psychologytoday.com/blog/flourish/200912/seeing-is-believing -the-power-visualization.

Page 75 *I had the privilege of consulting with my friend:* Pam Shriver, interview with the author, November 17, 2016.

Page 76 *just like in the climactic scene in the film:* Bernard Malamud, *The Natural* (1952; repr., New York: Farrar, Straus and Giroux, 2003).

Page 76 *He owes at least some of the credit:* George Mumford, *The Mindful Athlete: Secrets to Pure Performance* (Berkeley, CA: Parallax Press, 2016).

Page 77 *Billiards player Will DeYonker:* William DeYonker, "Right on Cue," You-Tube video, www.youtube.com/watch?v=FCNDCBE2lsE.

Page 78 *Multiple Olympic gold medal winner Kerri Walsh Jennings:* Sara Angle, "Olympic Beach Volleyball Player Kerri Walsh Jennings' Body Confidence Tips," *Shape*, March 11, 2015, www.shape.com/blogs/fit-famous/olympic -beach-volleyball-player-kerri-walsh-jennings-body-confidence-tips.

Page 79 *She won a whopping 112 career doubles titles:* "Pam Shriver," International Tennis Hall of Fame website, www.tennisfame.com/hall-of-famers /inductees/pam-shriver.

Page 80 *In fact, the 1927 Yankees' winning was such:* "1927: The Yankee Juggernaut," *This Great Game*, www.thisgreatgame.com/1927-baseball-history .html.

Page 80 *The University of Connecticut women's basketball team:* Geno Auriemma website, www.genoauriemma.com/geno/quotes.

Page 81 *the Republican refused to print the game box scores:* Larry Stone, "Think the UConn Women Are Too Good? Quit Whining and Beat 'Em," *Seattle Times*, March 29, 2016, www.seattletimes.com/sports/uw-husky -basketball/think-the-uconn-women-are-too-good-quit-whining -and-beat-em.

Page 81 *Among their claims to shame were:* "How Bad (and Lovable) Were the

1962 Mets?," *Jugs Sports*, jugssports.com/how-bad-and-lovable-were-the
-1962-mets.

Page 83 *Dubbed "Broadway Joe" for his stylish flair:* "Broadway Joe," YouTube
video, July 3, 2007, www.youtube.com/watch?v=Gc65NC44dSk.

Page 84 *Comedian Jerry Seinfeld has joked:* "We Cheer for Clothes," *Seinfeld*,
YouTube video, April 9, 2006, www.youtube.com/watch?v=we-L7w1K5Zo.

Page 86 *Perhaps the most tragic story of out-of-control:* "Murder of Soccer
Player after Own-Goal 20 Years Ago Still Resonates in Colombia,"
Fox News, July 2, 2014, www.foxnews.com/world/2014/07/02/murder
-soccer-player-after-own-goal-20-years-ago-still-resonates-in-colombia
.html.

Chapter 7. Immaculate Perception

Page 87 *In 2015, at New York City's popular:* Hannah Parry, "You Won't Brie-
Leave It: New York Restaurant Creates World's Most Expensive Grilled
Cheese Sandwich for $214," *Daily Mail*, June 10, 2015, www.dailymail
.co.uk/news/article-3118362/You-won-t-brie-leave-New-York-restaurant
-create-world-s-expensive-grilled-cheese-sandwich-214.html.

Page 87 *Serendipity's Quintessential Grilled Cheese Sandwich paled:* "'Virgin
Mary Grilled Cheese' Sells for $28,000," *NBC News*, November 23, 2004,
www.nbcnews.com/id/6511148/ns/us_news-weird_news/t/virgin-mary
-grilled-cheese-sells/#.WJzogbYrK1s.

Page 88 *In 2005 a home pregnancy test:* "Britney Spears' Pregnancy Test Sells,"
CNN, May 12, 2005, money.cnn.com/2005/05/12/news/newsmakers/britney
_pregnancytest.

Page 89 *Francis Bacon referred to it:* Francis Bacon, "The Plan of the Instauratio
Magna," www.bartleby.com/39/21.html.

Page 89 *I began the introduction to this book:* "Virgin Mary Seen in Tree Stump
in Limerick," *Belfast Telegraph*, September 7, 2009, www.belfasttelegraph
.co.uk/news/virgin-mary-seen-in-tree-stump-in-limerick-28486957.html.

Page 89 *On a pretzel that sold for $10,600:* "In a Twist of Fate — Holy Pretzel
Sells for $10,600," *PR Newswire*, June 2, 2005, www.prnewswire.com/news
-releases/in-a-twist-of-fate---holy-pretzel-sells-for-10600-54497527.html.

Page 89 *On a fence post near the cliffs:* "Apparition of Our Lady of Coogee
Beach," *Catholic News*, January 31, 2003, cathnews.acu.edu.au/301/166.php.

Page 89 *It has since aptly been named:* J. H. Crone, *Our Lady of the Fence Post*
(Perth, Aus: UWA Publishing, 2003).

Page 89 *Floating on a garage door in:* "Catholics Flock to Garage Door to See
Image of Virgin Mary," YouTube video, August 31, 2007, www.youtube.com
/watch?v=5jZld8Zg1aA.

Page 90 *As a pair of eyes on a bathroom door:* "Jesus on a Door," *Penn & Teller: Bullshit!*, YouTube video, September 3, 2014, www.youtube.com/watch ?v=ofCGV_zBEVo.

Page 90 *Faded in a window of Mercy Medical Center:* Bill Dusty, "Virgin Mary Apparition at Mercy Hospital," YouTube video, October 6, 2008, www .youtube.com/watch?v=L_g1YpkhvCA.

Page 90 *The Virgin Mary does get around:* Christopher Cihlar, *The Grilled Cheese Madonna and 99 Other of the Weirdest, Wackiest, Most Famous eBay Auctions Ever* (New York: Broadway, 2006).

Page 90 *including a now infamous appearance in some:* Joe Kovacs, "Jesus Appears in Shower, Worth $2,000," *WND*, June 25, 2005, www.wnd.com /2005/06/31018.

Page 90 *Jesus even found his way to the posterior:* "Image of Jesus on Dog's Butt God's Second Appearance?," *Patheos*, November 16, 2011, www.patheos .com/blogs/heavenlycreatures/2011/11/image-of-jesus-on-dogs-butt-gods -second-appearance.

Page 90 *Although devotees herald the blessings:* "Holy Grilled Cheese Sandwich! What Is Pareidolia?," *The Conversation*, theconversation.com/holy-grilled -cheese-sandwich-what-is-pareidolia-14170.

Page 90 *such as the "Paul is dead" hoax:* "The 'Paul Is Dead' Myth," *The Beatles Bible*, www.beatlesbible.com/features/paul-is-dead.

Page 90 *Nessie encounters:* Sarah Begley, "Loch Ness Monster Probably a Catfish, Says Man Who's Been Watching for 24 Years," *Time*, July 17, 2015, time.com/3962382/loch-ness-monster-catfish.

Page 91 *Let's take a minute to discuss why faces:* Naomi Greenaway, "What Do You See in These Photos?" *Daily Mail*, October 20, 2015, www.dailymail .co.uk/femail/article-3280816/What-photos-s-faces-suffer-facial-pareidolia .html.

Page 94 *Take the example of the little known:* Maureen A. Duffy, "Charles Bonnet Syndrome: Why Am I Having These Visual Hallucinations?," www.visionaware.org/info/your eye condition/guide-to-eye-conditions /charles-bonnet-syndrome/125.

Page 94 *For those stricken with CBS:* V. S. Ramachandran and Sandra Blakeslee, *Phantoms in the Brain: Probing the Mysteries of the Human Mind* (New York: William Morrow, 1999).

Page 95 *James Thurber, one of America's greatest:* James Thurber, "The Secret Life of Walter Mitty," *The Thurber Carnival* (1945; repr., New York: Harper-Collins, 2013).

Page 97 *Carl Sagan said it best:* Carl Sagan, *Cosmos* (New York: Ballantine, 2013).

Chapter 8. The Spell of the Sensuous

Page 99 *The Roman philosopher Cicero said:* Cicero, *Selected Works* (New York: Penguin, 1980).

Page 99 *The* reciprocation principle: Robert B. Cialdini, *Influence: The Psychology of Persuasion,* rev. ed. (New York: Harper Business, 2006), 140.

Page 102 *Ever since the nineteenth century:* The Basics of Philosophy, website, www.philosophybasics.com/branch_altruism.html.

Page 102 *A 2006 study involving toddlers:* Bjorn Carey, "Stanford Psychologists Show That Altruism Is Not Simply Innate," *Stanford Report,* December 18, 2014, news.stanford.edu/pr/2014/pr-altruism-triggers-innate-121814.html.

Page 109 *When a man is threatened his:* "New Vision on Amygdala after Study on Testosterone and Fear," *Science Daily,* June 12, 2015, www.sciencedaily.com/releases/2015/06/150612143027.htm.

Page 109 *In one scientific study, when women were injected:* Erno J. Hermans et al., "A Single Administration of Testosterone Reduces Fear-Potentiated Startle in Humans," *Biological Psychiatry* 59, no. 9 (June 2006): 872–74, www.researchgate.net/publication/7316503_A_Single_Administration_of _Testosterone_Reduces_Fear-Potentiated_Startle_in_Humans.

Page 109 *The origin of this perception goes back to:* William Congreve, *The Mourning Bride* (London: Dodo Press, 2008).

Page 109 *Back in the early 1990s, a scandal:* Darren Boyle, "I Didn't Know She Cut It Off: Penis Attack Victim John Bobbitt Reveals the Horror of Being Assaulted by His Wife in Notorious Crime," *Daily Mail,* November 24, 2016, www.dailymail.co.uk/news/article-3968154/I-didn-t-know-cut-Penis -attack-victim-John-Bobbitt-reveals-horror-assaulted-wife-notorious -crime.html.

Chapter 9. Star Time

Page 112 *Years ago Hall of Fame baseball player:* "The Best Celebrities You'd Ever Buy Stuff From," *Ranker,* www.ranker.com/list/the-best-celebrity -tv-spokespeople-of-all-time/josh-heller.

Page 113 *In the past, TV personality Jenny McCarthy:* Dina Fine Maron, "Sorry, Jenny McCarthy: Vaccines Aren't as Dangerous as You Think," *Salon,* January 11, 2017, www.salon.com/2017/01/11/sorry-jenny-mccarthy-vaccines -arent-as-dangerous-as-you-think_partner.

Page 113 *Kim Kardashian West receives:* Laura Case, "How Much Do Celebrities Get Paid to Tweet?" *Wet Paint,* July 20, 2016, www.wetpaint.com/how -much-do-celebrities-get-paid-to-tweet-663232.

Page 119 *Ms. Kardashian West, who has been dubbed:* Kim Kardashian West, *Kim Kardashian Selfish* (New York: Universe, 2015).

Chapter 10. Persexual Intelligence 101

Page 122 *If you're among the few human beings:* "2014 Survey· How Many Christians Do You Think Watch Porn?," *Digital Journal,* press release, August 14, 2014, www.digitaljournal.com/pr/2123093.

Page 122 *30 percent of all Internet* "30 Percent of Global Web Traffic Is Porn – Study," Indo-Asian News Service, June 5, 2012, http://gadgets.ndtv.com /internet/news/30-percent-of-global-web-traffic-is-porn-study-223878.

Page 122 *One out of every three women:* "Study Finds That 1 Out of 3 Women Watch Porn at Least Once a Week," Women in the World feature, *New York Times,* October 22, 2015, nytlive.nytimes.com/womeninthe world/2015/10/22/study-finds-that-1-out-of-3-women-watch-porn-at -least-once-a-week.

Page 123 *not only do the primary remains of the baths:* Toby Meyjes, "Up Pompeii! Erotic Paintings Reveal Sex Lives of Ancient Romans," *Metro,* December 8, 2016, metro.co.uk/2016/12/08/up-pompeii-erotic-paintings -reveal-sex-lives-of-ancient-romans-6308999.

Page 124 *Somewhere around the second century:* Indra Sinha, *The Love Teachings of Kama Sutra* (New York: Marlow, 1997).

Page 125 *To this day, many people remain convinced:* Marissa Fessenden, "Medieval Chastity Belts Are a Myth," *Smithsonian,* August 20, 2015, www.smithsonianmag.com/smart-news/medieval-chastity-belts-are -myth-180956341.

Page 125 *Chastity belts didn't come into vogue:* Radhika Sanghani, "Chastity Belts: The Odd Truth about 'Locking Up' Women's Genitalia," *Telegraph,* January 18, 2016, www.telegraph.co.uk/women/sex/chastity-belts-the-odd -truth-about-locking-up-womens-genitalia.

Page 125 *Ironically, the opposite is true:* Marc Garnick, "Does Frequent Ejaculation Help Ward Off Prostate Cancer?," *Prostate Knowledge,* Harvard Medical School publication, www.harvardprostateknowledge.org/does frequent-ejaculation-help-ward-off-prostate-cancer.

Page 126 *From the 1960s to the 1980s:* "Rev Morton Hill, 68, Pornography Opponent," *Chicago Tribune,* obituary, November 7, 1985.

Page 126 *Father Hill was by no means alone:* Richard N. Ostling, "Jerry Falwell's Crusade," *Time,* June 24, 2001, content.time.com/time/magazine/article /0,9171,142305,00.html.

Page 128 *Not much has changed since the classic 1969 novel:* Philip Roth, *Portnoy's Complaint* (1969; repr., New York: Vintage, 1994).

Page 129 *six out of ten girls have been exposed:* "Pornography Statistics: 2015 Report," Covenant Eyes, www.covenanteyes.com/pornstats/.

Page 129 *In 1879 an internationally famous author:* Mark Twain, *Mark Twain on Masturbation: Some Thoughts on the Science of Onanism* (CreateSpace, 2009).

Page 131 *A staggering 41 percent of couples:* "Infidelity Statistics," *Infidelity Facts,* website, www.infidelityfacts.com/infidelity-statistics.html.

Page 131 *raked in a whopping $119 million globally:* Box Office Mojo website, www.boxofficemojo.com/movies/?id=unfaithful.htm.

Page 132 *One can understand, given the extreme reactions:* Chris Spargo, "How Rock Hudson Kept His Gay Life Secret…," *Daily Mail,* April 15, 2015, www.dailymail.co.uk/news/article-3040119/Secret-gay-life-Rock-Hudson -revealed-man-called-one-true-love-describes-pair-hid-relationship -protect-image-Hollywood-star.html.

Page 133 *Many other stars believed to have been gay:* Claude J. Summers, ed., *The Queer Encyclopedia of Film and Television* (San Francisco: Cleis, 2015).

Page 133 *the books* Full Service: Scotty Bowers, *Full Service: My Adventures in Hollywood and the Secret Sex Lives of the Stars* (New York: Grove, 2012); Boze Hadleigh, *In or Out: Gay and Straight Celebrities Talk about Themselves and Each Other* (Fort Lee, NJ: Barricade, 2000).

Page 133 *More than a century ago, the brilliant:* The Official Site of Oscar Wilde, website, www.cmgww.com/historic/wilde.

Page 133 *and we have these and many other megatalents:* "100 Actors Who Are Actually Gay or Bisexual," *IMDb,* www.imdb.com/list/ls072706884.

Page 134 *In July 2015* Vanity Fair *published:* Buzz Bissinger, "Caitlyn Jenner: The Full Story," *Vanity Fair,* July 25, 2015, www.vanityfair.com/hollywood /2015/06/caitlyn-jenner-bruce-cover-annie-leibovitz.

Page 135 *In Val Camonica, Italy:* Emmanuel Anati, "The Way of Life Recorded in the Rock Art of Valcomonica," *Adoranten 2008,* www.rockartscandinavia .com/images/articles/a8marretta.pdf.

Page 135 *In the 1950s sex researcher Dr. Alfred Kinsey:* "Bestiality Is Much, Much More Common Than You Think," February 20, 2015, *Health24,* website, www.health24.com/sex/sexual-diversity/bestiality-is-much-much-more -common-than-you-think-20150218.

Page 136 *Denmark recently passed laws against:* "Denmark Passes Law to Ban Bestiality," *Newsbeat,* BBC publication, April 22. 2015, www.bbc.co.uk /newsbeat/article/32411241/denmark-passes-law-to-ban-bestiality.

Page 136 *The United States is all over the place:* Robert Emmett Murphy Jr., "Bestiality Is Legal in the Same States That Ban Same-Sex Marriage,"

All Things Crime, blog, July 16, 2013, allthingscrimeblog.com/2013/07/16
/bestiality-is-legal-in-the-same-states-that-ban-same-sex-marriage.

Chapter 11. Gotta Have It

Page 139 *At a London coffeehouse, people pay:* "All About Kopi Luwak — The
Most Expensive Coffee in the World!," www.most-expensive.coffee.

Page 140 *An odd nocturnal Asian animal:* "Civet Coffee: Why It's Time to Cut
the Crap, *The Guardian*, September 13, 2013, www.theguardian.com/life
andstyle/wordofmouth/2013/sep/13/civet-coffee-cut-the-crap.

Page 140 *sometimes mistreated creatures:* Rachael Bale, "The Disturbing Secret
behind the World's Most Expensive Coffee," *National Geographic*, April
29, 2016, news.nationalgeographic.com/2016/04/160429-kopi-luwak
-captive-civet-coffee-Indonesia.

Page 141 *Many will remember the intense fervors:* Stephanie Buck, "The Weird,
Rabid History of the Cabbage Patch Craze," *Timeline*, December 14, 2016,
timeline.com/cabbage-patch-craze-867ce8d076c#.2ohk9bbh6

Page 141 *What made these items seem so essential:* Larry Getlen, "How the
Beanie Baby Craze Was Concocted — Then Crashed," *New York Post*,
February 22, 2015, nypost.com/2015/02/22/how-the-beanie-baby-craze
-was-concocted-then-crashed.

Page 144 *Right there, for only a buck and a quarter:* Erin McCarthy, "16 Amazing
Facts about Sea Monkeys," *Mental Floss*, mentalfloss.com/article/56755/16
-amazing-facts-about-sea-monkeys

Page 145 *Delusions of grandeur were not the only:* "A Fortune from Sea Monkeys and X-Ray Spex: Harold von Braunhut, Novelty Gift Entrepreneur,
1926–2003," *Sidney Morning Herald*, obituary, www.smh.com.au/articles
/2004/01/01/1072908849191.html.

Page 146 *Some of these alluring objects:* Kara Kovalchik, "11 Shameless Comic Book
Ads That Cost Us Our Allowance," *Mental Floss*, mentalfloss.com/article
/30420/11-shameless-comic-book-ads-cost-us-our-allowance-money.

Page 146 *More recently, cheesy noisemaking gag gifts:* Catherine Greenman, "A
Singing Fish Gets a Personal Touch," *New York Times*, January 11, 2001,
www.nytimes.com/2001/01/11/technology/a-singing-fish-gets-a-personal
-touch.html.

Chapter 12. Are You Different from a Wildebeest in Kenya?

Page 152 *In his book* Influence, *Robert Cialdini asserts:* Robert B. Cialdini, *Influence: The Psychology of Persuasion*, rev. ed. (New York: Harper Business,
2006), 171–72.

Page 153 *In a 2014 study, "Social Defaults:* Young Eun Huh, Joachim Vosgerau, and Carey K. Morewedge, "Social Defaults: Observed Choices Become Choice Defaults," *Journal of Consumer Research* 41, no. 3 (October 2014), www.jstor.org/stable/10.1086/677315?origin=JSTOR-pdf&seq=1#page _scan_tab_contents.

Page 154 *If you are of the baby boomer generation,* "Woodstock 1969," *Woodstock Story,* www.woodstockstory.com/woodstock1969.html.

Page 155 *Flash-forward thirty years:* Daniel Kreps, "19 Worst Things about Woodstock '99," *Rolling Stone,* July 31, 2014, www.rollingstone.com/music /news/19-worst-things-about-woodstock-99-20140731.

Page 157 *We're gonna try to do a sing-along:* "Sly Stone's Speech at Woodstock," YouTube video, October 6, 2008, www.youtube.com/watch?v=lUr5rzQZkVg.

Page 157 *By contrast, the 1999 event:* Jane Ganahl, "Woodstock '99: The Day the Music Died," *SFGate,* July 28, 1999, www.sfgate.com/style/article /Woodstock-99-The-day-the-music-died-3073934.php.

Page 159 *Before the turn of the millennium:* "11 Celebrities Who Got Scammed by Bernie Madoff and Lost Millions," *Fox Business,* February 2, 2016, www.foxbusiness.com/features/2016/02/02/11-celebrities-who-got-scammed -by-bernie-madoff-and-lost-millions.html.

Page 159 *Psychologists and social scientists believe:* Robert Lenzner, "Bernie Madoff's $50 Billion Ponzi Scheme," *Forbes,* December 12, 2008, www.forbes.com/2008/12/12/madoff-ponzi-hedge-pf-ii-in_rl_1212croesus _inl.html.

Page 159 *who blindly fell for the facade of* social proof: Robert B. Cialdini, *Influence: The Psychology of Persuasion,* rev. ed. (New York: Harper Business, 2006), chap. 4.

Page 160 *Consider the following examples from years past:* Will Heilpern, "18 False Advertising Scandals That Cost Some Brands Millions," *Business Insider,* March 31, 2016, www.businessinsider.com/false-advertising-scandals -2016-3/#vw-falsely-advertised-environmentally-friendly-diesel-cars-1.

Page 161 *Millions of people bought the aforementioned:* "The Too Good to Be True Product Hall of Fame," *Time,* October 6, 2011, business.time.com /2011/10/11/14-products-with-notoriously-misleading-advertising -claims/slide/splenda.

Page 161 *I do not mean to suggest that:* Elaine Watson, "Kashi Agrees to Pay up to $3.99M to Settle 'All Natural' Lawsuit; Campbell Soup under Fire over Prego Labels," *Food Navigator,* William Reed newsletter, June 15, 2015, www.foodnavigator-usa.com/Regulation/Kashi-agrees-to-pay-up-to-3.99m -to-settle-all-natural-lawsuit.

Page 162 *Tom Hanks has ranked high:* Czarina Ong, "Tom Hanks Is the Most Trusted Celebrity While NBC's Brian Williams Took a Nosedive in

Ratings," *Christian Today*, February 12, 2015, www.christiantoday.com
/article/tom.hanks.is.the.most.trusted.celebrity.while.nbcs.brian.williams
.took.a.nosedive.in.trust.ratings/47947.htm.

Page 163 *When it comes to the life span of a battery:* Noreen O'Leary, "Dannon
Settles False Ad Suit for $35 Mil," *Adweek*, September 18, 2009, www.adweek
.com/brand-marketing/dannon-settles-false-ad-suit-35-mil-106416.

Chapter 13. Fanaticism

Page 165 *Fired-up antiwar demonstrators:* "San Francisco Protestors Stage a
'Vomit In,'" *SFGate*, www.sfgate.com/news/article/San-Francisco
-protesters-stage-a-vomit-in-2627438.php.

Page 167 *There is the harrowing tale:* Elizabeth Burchard and Judith L. Carlone,
The Cult Next Door: A True Story of a Suburban Manhattan New Age Cult
(San Mateo, CA: Ace Academics, 2011).

Page 168 *The answer, of course, is God:* "Pope Urban II Orders First Crusade,"
History, www.history.com/this-day-in-history/pope-urban-ii-orders-first
-crusade.

Page 169 *He would have no inkling:* "How Many People Died in the Crusades?,"
Reference, www.reference.com/history/many-people-died-crusades
-4483019b5f8684c5.

Page 170 *One recent example, the Second Sudanese Civil War:* "Sudan Referen-
dum: Key Dates in Recent History," *Telegraph*, January 8, 2011, www
.telegraph.co.uk/news/worldnews/africaandindianocean/sudan/8246617
/Sudan-referendum-key-dates-in-recent-history.html.

Page 170 *In many countries today:* Drew DeSilver and David Masci, "World
Muslim Population More Widespread Than You Might Think," *Fact Tank*,
Pew Research Center publication, January 31, 2017.

Page 171 *Dr. Marvin Galper, a San Diego–based:* Dr. Marvin Galper, interview
with the author, December 11, 2016.

Page 172 *What Do Tom Cruise, Kelly Preston, John Travolta:* Andy Cush, "A
Comprehensive List of Every Celebrity Connected to Scientology,"
Gawker, April 2, 2015, gawker.com/a-comprehensive-updated-list-of
every celebrity linked 1604554376

Page 172 *Scientology was the brainchild of L. Ron Hubbard:* Jason Guerrasio,
"The Chilling Story of How Scientology Founder L. Ron Hubbard Rose
to Power," *Business Insider*, March 31, 2015, www.businessinsider.com/l
-ron-hubbard-history-sci-fi-writer-to-scientology-founder-2015-3.

Page 172 *In 1950 he assembled his unique thoughts:* L. Ron Hubbard, *Dianetics:
The Modern Science of Mental Health* (1950; repr., Commerce, CA: Bridge
Publications, 2007).

Page 173 *Although Scientology has some superficial:* "What Is Scientology and Who Was L. Ron Hubbard?," *Telegraph*, October 6, 2016, www.telegraph .co.uk/news/0/what-is-scientology-and-who-was-l-ron-hubbard.

Page 173 *I recommend that you read the works of:* Norman Vincent Peale, *The Power of Positive Thinking* (1952; repr., Important Books, 2013).

Page 173 *In 1978 Jim Jones, founder:* "Jonestown," *History*, www.history.com /topics/jonestown

Page 173 *Believe it or not, this cult still exists:* Michael Zennie, "New Age Followers Still Waiting for Aliens to Beam Them Up 15 Years after Heaven's Gate Cult Suicides Left 39 People Dead," *Daily Mail*, March 26, 2012, www.dailymail.co.uk/news/article-2120869/Heavens-Gate-cult-committed -mass-suicide-15-years-ago.html.

Chapter 14. The Subjective Experience of Time

Page 177 *In a 2016 study published:* "Humans Perceive Time Somewhere between Reality and Our Expectations," University of Birmingham website, July 14, 2016, www.birmingham.ac.uk/news/latest/2016/07/humans-perceive -time-somewhere-in-between-reality-and-our-expectations.aspx.

Page 177 *Researchers at the San Jose Faculty of Medicine:* Cheyenne MacDonald, "Take the Test That Reveals How You Perceive Time," *Daily Mail*, July 13, 2016, www.dailymail.co.uk/sciencetech/article-3688988/Is-perception -time-wrong-Study-finds-humans-perceive-time-expectation-reality.html.

Page 179 *A toddler will not view time:* Penelope Leach, "Why Is My Toddler So Impatient?," *Baby Centre*, Babycentre.com, www.babycentre.co.uk /x539875/why-is-my-toddler-so-impatient.

Page 181 *When we attempt to swat a fly:* Emilie Reas, "Small Animals Live in a Slow-Motion World," *Scientific American*, July 1, 2014, www.scientific american.com/article/small-animals-live-in-a-slow-motion-world.

Page 181 *many domesticated dogs, according to Animal Planet:* Animal Planet, *It's Me or the Dog*, see "Dogs with Separation Anxiety, Ask Victoria Stilwell," www.animalplanet.com/tv-shows/its-me-or-dog/training-tips/separation -anxiety.

Page 181 *But imagine if I lived in 1907:* "Life Expectancy in the USA, 1900–98," u.demog.berkeley.edu/~andrew/1918/figure2.html.

Page 181 *If you're lucky enough to live in Monaco:* "The World Factbook," CIA publication, www.cia.gov/library/publications/the-world-factbook /rankorder/2102rank.html.

Page 184 *According to a 2015 study:* Stacy Liberatore, "Can You Believe It's Already Christmas? Technology Is SPEEDING UP Our Perception of Time,

Researchers Say," *Daily Mail*, November 19, 2015, www.dailymail.co.uk
/sciencetech/article-3325763/Can-t-believe-s-Christmas-Technology
-SPEEDING-perception-time-claims-study.html.

Chapter 15. Gut Check

Page 188 *On February 3, 1959, rock stars:* Greil Marcus, "Ritchie Valens, J. P. 'The
Big Bopper' Richardson, and Buddy Holly," *Rolling Stone*, June 28, 1969,
www.rollingstone.com/music/news/ritchie-valens-j-p-the-big-bopper
-richardson-and-buddy-holly-19690628.

Page 188 *What is not quite as well known is that:* Dave Lifton, "The Story of
Rock's First Tragedy: Buddy Holly, Ritchie Valens and the Big Bopper
Killed in Plane Crash," Ultimate Classic Rock, website, February 3, 2016,
ultimateclassicrock.com/buddy-holly-richie-valens-big-bopper-killed
-in-plane-crash.

Page 189 *What intrigues me most about this story:* R. Gary Patterson, *Take a
Walk on the Dark Side: Rock and Roll Myths, Legends, and Curses* (New
York: Touchstone, 2004), chap. 2.

Page 190 *Edison inventing the phonograph:* "Thomas Edison, Genius Inventor,
Dies at 84," *New York Daily News*, October 17, 2015 (originally published
October 19, 1931), www.nydailynews.com/news/national/thomas-edison
-genius-celebrated-funeral-1931-article-1.2389894; *Biography*, s.v. "Nikola
Tesla," accessed May 21, 2017, www.biography.com/people/nikola
-tesla-9504443.

Page 190 *The CFA — an organization devoted to:* Jason Voss, "The Intuitive
Investor: A Simple Model of Intuition," *Enterprising Investor*, blog for CFA
Institute, blogs.cfainstitute.org/investor/2014/07/22/the-intuitive-investor
-a-simple-model-of-intuition.

Page 192 *and other creatures who use the Earth's:* Tricia Edgar, "Animal Mag-
netism: How the Magnetic Field Influences Animal Migration," *Decoded
Science*, November 30, 2014, www.decodedscience.org/animal-magnetism
-magnetic-field-influences-animal-navigation/50745.

Page 192 *and dogs with schnozzes so powerful they:* Elizabeth Cohen and John
Bonifield, "Meet the Dogs Who Can Sniff Out Cancer Better Than Some
Lab Tests," *Vital Signs*, CNN publication, February 3, 2016, www.cnn.com
/2015/11/20/health/cancer-smelling-dogs.

Page 193 *Numerous scientific studies have shown that:* Leonardo Vintini, "Intuition: The Sense That Defies the Physical," *Epoch Times*, January 19, 2009,
www.theepochtimes.com/n3/1526638-intuition-dr-alan-pegna-tsunami
-ronald-rensink-ancient-culture.

Page 194 *One of the pioneers in the field of intuition:* Robert Lee Hotz, "A Pioneer in the Private Life of the Brain," *Wall Street Journal*, February 11, 2011, www.wsj.com/articles/SB10001424052748704364004576132262901047364.

Page 194 *His study involved sixteen participants:* Antoine Bechara, Hanna Damasio, Antonio R. Damasio, Gregory P. Lee, et al., "Different Contributions of the Human Amygdala and Ventromedial Prefrontal Cortex to Decision-Making," *Journal of Neuroscience*, July 1, 1999, www.jneurosci.org /content/jneuro/19/13/5473.full.pdf.

Page 194 *Scientists Galang Lufityanto:* "Intuition — It's More Than a Feeling," *APS*, publication for the Association for Psychological Science, April 21, 2016, www.psychologicalscience.org/news/minds-business/intuition-its -more-than-a-feeling.html#.WJ8N9rYrKb8.

Page 195 *In his groundbreaking book* Blink: Malcolm Gladwell, *Blink: The Power of Thinking without Thinking* (New York: Little, Brown, 2005), 174–75.

Page 196 *For the late John Lennon:* "Seven Things You Didn't Know About 'Lucy in the Sky with Diamonds,'" *OUPblog*, Oxford University Press blog, July 9, 2013, blog.oup.com/2013/07/lucy-in-the-sky-with-diamonds-facts.

Chapter 16. Your PI Assessment

Page 215 *In his benchmark work* Emotional Intelligence: Daniel Goleman, *Emotional Intelligence: Why It Can Matter More Than IQ* (New York: Bantam, 2005).

Index

acceptance, 93
ACE Family (YouTube channel), 119
Activia yogurt, 160, 162–63
Addicted (film; 2014), 131
advertising, 122, 144, 146–47, 153, 160–63, 202
Affair, The (TV series), 131
Afghanistan, 170
afterlife, 45–46, 51–52
age, and time perception, 181
aggression, 109
AIDS, 132
Aldrin, Buzz, 178
Alexander, Eben, 47–48
Alexios I Komnenos (Byzantine emperor), 169
alien sightings, 33–34, 35
Allen, Woody, 45
Alley, Kirstie, 172
allies, 103–4
All in the Family (TV show), 179, 195
Allsup, Tommy, 188
Allyson, June, 112
alternative medicine, 23
altruism, 102–3
Alzheimer's disease, 72
American Pie (film; 1999), 128
"American Pie" (McLean), 188
amputees, 27
amygdala, 109
Andersen, Hans Christian, 28
Animal Planet (TV channel), 181

animal-rights fanaticism, 165
animals: intuition of, 192–93; perception differences of, 5–6; sexual acts with, 135–37; time as perceived by, 181
Aniston, Jennifer, 117
antismoking fanaticism, 165
antivaccination movement, 113
antiwar fanaticism, 165
anxiety disorder, 28
apologies, 110
Applewhite, Marshall, 173
Archer, Anne, 172
Aristotle, 18
Armstrong, Lance, 82, 83–84
Armstrong, Neil, 178
Arsenio Hall Show, The (TV talk show), 67
art, 37–41, 126
Auriemma, Geno, 81
autism, 77, 113

Bachelor, The (reality TV show), 118
Bacon, Francis, 89
Bad News Bears films, 81
Baker, Josephine, 133
Baldwin, Alec, 114
Baltimore Colts, 82
Barnum, P. T., 159
Barrel of Monkeys, 160n
Bartman, Steve, 86
bathroom controversies, 135
Beanie Babies, 141
Beatles, 97, 190, 195, 196

Beauregard, Mario, 45
Beck (musician), 172
behavioral mimicry, 153
Being Human Conference (2012), 43
Being Mary Jane (TV series), 131
Bell, Drake, 134
Bennett, Tony, 54
Berkeley, George, 14
Berle, Milton, 116
Berra, Yogi, 85
Best, George, 88
bestiality, 135–37
Beyoncé, 114
biases, 96
Bible, 124–25
Big Bopper (J. P. Richardson), 188–90, 200
Biggs, Jason, 128
Big Mouth Billy Bass, 146
bisexuality, 133
Blaine, David, 31
blindness, cortical, 57
blindsight, 57–58
Blink (Gladwell), 195
Bobbitt, John (Wayne), 109, 110
Bobbitt, Lorena, 109
Bodenhamer, Bob G., 6
body, and brain, 12
bondage, 129
Boston Red Sox, 84, 85
Bowers, Scotty, 133
Boxer Wachler, Brian: contact info,
    220; family name of, 108n; intuitive
    experiences of, 191–92; OBE of, 55–56;
    as ophthalmologist, 2–3, 41–42, 42n,
    59, 78–79, 119; PI of, 78–79; at Sochi
    Olympics (2014), 59; TEDx Talk given
    by, 42n; teen fixations of, 184–85
Boxer Wachler, Selina, 108n, 191–92, 192n
brain: art and, 37–38; during clinical
    death, 49–50; consciousness outside
    of, 46; critical thinking and, 213; eyes
    and, 95; gray matter in, 72; healing
    power of, 18, 19; illusion and, 42–43;
    immaculate perception and, 96;
    interconnections of, 11–12; during
    NDEs, 51–52; perception changed by,
    2–4; physical exercise and, 70–73, 76;
    research on, 18; revenge and, 107, 109;
    senses tricked by, 27; sensory process
    in, 15–16, 18–19; white matter in, 72–73,
    76
BrainDecoder.com, 16

brain injuries, 57
Brain Research through Advancing In-
    novative Neurotechnologies (BRAIN)
    Initiative, 18
brainwashing, 167–68, 171–72
branding, 163n
Braunhut, Harold von, 145
Brimley, Wilford, 112, 113
Brown, Derren, 31
Bruce, Lenny, 115–16
Bryant, Kobe, 82
Brynner, Yul, 130
bucket lists, 177–78
Burchard, Elizabeth R., 167
Burpo, Colton, 47, 52–53
Burr, Raymond, 133
Busey, Gary, 54
Bush, George W., 66
business, 103–4
buyer's remorse, 143–44

Cabbage Patch dolls, 141
Caligula (film; 1979), 142n
Callow, Simon, 133
cancer, 22–23
car dealerships, reciprocity games of, 100,
    101–2, 199
cardiac arrest, 49–50, 55
cardiovascular disease, 23
Carlin, George, 116
Carr, Kris, 22–23, 26
Carroll, Lewis, 196
Cathedral of Christ the Savior (Moscow,
    Russia), 62
Catherine the Great (Empress of Russia),
    136
Catholicism, 50–51, 89, 92
cat poop coffee, 139–40
Cavuto, Neil, 134
celebrities: as activists, 212–13; comedians
    as, 115–17; critical thinking regarding,
    212–13; as cult members, 172; defined,
    111; gay/lesbian/bisexual, 133; as hypo-
    chondriacs, 28; with MS, 22; NDEs of,
    53–54; PI influenced by (halo effect),
    111–15, 116–17, 202; product endorse-
    ments by, 162–63, 200, 202, 211; reality
    TV and, 118–19; "selfie stars," 119–20;
    social media and, 118–20; stalkers of,
    114–15; tabloid reporting on, 117, 118
celebrity fans, 119–20
censorship, 126, 142, 142n

cerebral neocortex, 16
Cetewayo (Zulu hero), 130
CFA, 190–91
Chad, 181
Charles Bonnet Syndrome, 3, 94–95
Chase, Chevy, 54
chastity belts, 125–26
Chattanooga (TN), 35
Chicago Bulls, 80
Chicago Cubs, 85, 86
child pornography, 121
children: altruistic behavior of, 102–3;
    time perception and, 181
Chimelis, Ron, 81
Cho, Margaret, 116
Christianity, 168–70
Christmas gifts, 142–43
chromostereopsis, 38
chuckle fuckers (comedy groupies), 117
Church of Scientology, 172–73
Cialdini, Robert, 99–100, 104, 152–53
Cicero, 99
civet, Asian palm, 140
Clift, Montgomery, 133
climate change, 178, 200, 204–5
Clinton, Bill, 54, 67
Clinton, Chelsea, 54
Clinton, Hillary, 54
Clooney, George, 207, 212–13
Close, Chuck, 38
CNN, 134
coffee, cat poop (Kopi Luwak), 139–40
color, 38
comedians, 115–17
comedy clubs, 116–17
comic books, items sold in, 144–45
commercialism, 157–58
complaining, 26
compromise, 103–4
Comte, August, 102
concessions, 103–4
conformity, 152, 171–72
Congreve, William, 109
connection, 79, 80
consciousness: after death, 50–51; neural
    correlates of, 49; outside the brain, 46
"Conversations with Helmholtz" (Allen),
    45
Copperfield, David, 31
cortical blindness, 57
Costello, Carol, 134
Crawford, Jamal, 26

Crazy Sexy Cancer Tips (Carr), 23
Crichton, Michael, 130
critical thinking, 212–13
Crone, J.H., 89
Cruise, Tom, 172
Crusades, 168–70, 209, 215
cults, 167, 170–74, 200, 201, 212
culture: NDEs and, 50–51, 52–53; time
    perception and, 181
Curtis, Jamie Lee, 162–63, 200
Cusack, John, 114

Damasio, Antonio R., 194
dark lucid dreams, 33–36, 37, 90, 199, 212
"Day in the Life, A" (Beatles), 196
death: consciousness following, 50–51;
    determining time of, 48–49; OBEs
    during, 55–56; premonitions of,
    188–90; subjective nature of, 48. See
    also near-death experiences (NDEs)
Deep Throat (film; 1972), 142n
Defense Intelligence Agency, 32
DeGeneres, Ellen, 133, 179
dementia, 72
De Niro, Robert, 114–15
Denmark, 136
Derman, Emanuel, 191
DeYonker, Will, 77
Diagnostic and Statistical Manual of
    Mental Disorders (DSM), 27
Dianetics (Hubbard), 172
Didion, Joan, 22
Dietrich, Marlene, 133
Di Luca, Max, 177
DiMaggio, Joe, 112
DiMucci, Dion, 188
discovery, 93
discrimination, 178–79
Divorce (TV series), 131
dogs, time perception of, 181
Donkin, Chris, 194
dopamine, 197, 199, 143, 177
doping, 83–84
Double Indemnity (film; 1944), 131
doubt, 93
dreams, 202–3. See also dark lucid dreams
dry eye disease, 25–26
Duyser, Diana, 87–88
Dylan, Bob, 196

eBay, 88
Edinburgh University (Scotland), 55–56

Eeden, Frederik van, 34
Einstein, Albert, 1, 2, 190
electromagnetic waves, 19
electronic devices, 184–85
Elvis sightings, 90
email spam, 174
Emotional Intelligence (Goleman), 215
emotions, focusing on, 215–17
"Emperor's New Clothes, The"
    (Andersen), 28
endorphins, 70, 71
Enterprising Investor (CFA publication),
    191
Ephesus (ancient Greek city), 123
Escobar Saldarriaga, Andrés, 86
estrogen, 108–9
eugenics, 125n
Europe, brain research in, 18
Evidence of the Afterlife (Long), 46
evolution, 178
exercise, physical: brain influenced by,
    70–73, 76; PI and, 69–70; procrastina-
    tion regarding, 71
Expressionist painting, 40–41
ExtenZe, 160
extra-corporeal experiences (ECEs),
    56. *See also* out-of-body experiences
    (OBEs)
eye freckles, 26
eyesight, and brain, 2–3

Facebook, 29, 99, 203–4
faces, 91–92
faith healers, 174
fake news, 203–4, 213–14
fanaticism: cults, 170–74, 200, 201, 212; PI
    and, 165–68, 169, 171–72, 174, 200, 201;
    religion-inspired, 165, 168–71; of sports
    fans, 68, 70, 84–86, 111, 166, 199, 201;
    types of, 165
Farrell, Colin, 114
Fatal Attraction (film; 1987), 109, 131
fear, 24, 34–35, 104, 109
feminist organizations, 127
Finland, 136
Flaubert, Gustave, 126
fMRI technology, 42
focus, 75
Food and Drug Administration (FDA),
    161
forgiveness, 24
Foster, Jodie, 133

Fox News, 134
Franklin, Aretha, 97
Freud, Sigmund, 127
Friedman, Kinky, 134
Frontiers in Human Neuroscience, 56
Full Service (Bowers), 133

gag gifts, 144–45
Gallup polls, 46
Galper, Marvin, 171, 172, 174
Gandhi, Mahatma, 97
Garbo, Greta, 133
Garlington, Lee, 132–33
Garr, Teri, 22
gay marriage, 134
Gehrig, Lou, 85
Geller, Uri, 32
gender identity, 134–35
Gere, Richard, 131
Gholipour, Bahar, 16
ghosts, 36
gifts/gift giving, 99–100, 105–7, 142–43
Gladwell, Malcolm, 195
Gleason, Jackie, 116
gold, perceived value of, 140–41
Goleman, Daniel, 215
Gomez, Selena, 114
Good Wife, The (TV series), 131
Gore, Al, 204–5, 215
Grant, Cary, 160
Greeks, ancient, 123, 124
Green Bay Packers, 84
grilled cheese sandwich, Virgin Mary
    image on, 87–88, 140, 199
Groundhog Day (film; 1993), 183–84
grudge holding, 110
Guatemala, IOC meeting in, 60
guilt, 97, 101–2, 104

Hadleigh, Boze, 133
Hale-Bopp Comet, 173
Hall, L. Michael, 6
hallucinations, 1, 3, 27, 49–50, 94–95
halo effect, 111–15, 116–17, 200
Hamilton, Laird, 77–78
Hangover, The (film; 2009), 83
Hanks, Tom, 162, 162n, 202, 211
Harris, Neil Patrick, 133
Harris, Sam, 11, 46
Harrison, George, 114
Hayes, Isaac, 172
health scares, 29

heart attacks, 23
Heaven Is for Real (Burpo), 47
Heaven's Gate (cult), 173
Hepburn, Katharine, 133
herd mentality, 28, 151, 153, 213
"Higher" (Sly and the Family Stone), 156–57
Hill, Morton A., 126
Hinduism, 124
hippie culture, 156
Hitchcock, Alfred, 160
Hitler, Adolf, 169
Hodges, Gil, 82
Hoffman, Abbie, 154–55
Holcomb, Steven, 41, 59, 64, 78
Holcomb C3-R® (eye treatment), 41–42, 119
Holly, Buddy, 188, 189, 200
Holly, Maria Elena, 189
Homer, 130
homophobia, 178–79
homosexuality, 132–34, 178–79
Hook, Jay, 81
Houdini, Harry, 31
How to Get Away with Murder (TV series), 131
Hubbard, L. Ron, 172–73, 174
Hudson, Rock, 132–33
hypochondria, 27–29, 199

I Am Curious Yellow (film; 1967), 142n
Iliad, The (Homer), 130
illness perception, xvii–xix, 22–29
illusion: art and, 37–41; as dark lucid dreams, 33–36, 199; defined, 26–27; of godlike power, and spending habits, 144–45; human fascination with, 31–33, 96–97; hypochondria and, 27–29, 199; PI and, 27, 31, 42–43, 46; psychic, 32; reality as, 1, 2; scientific research on, 41–43; unexplained mysteries and, 36–37
imagination, 93, 97, 220
immaculate perception: examples of, 87–90; pareidolia as cause of, 90–92; PI and, 96–97, 203
immediacy, 13
immune system, 24
Impressionist painting, 38–39
incest, 121
India, ancient, 124
individual thinking, 213–15
infidelity, 121–22, 130–32

Influence (Cialdini), 99–100, 104, 152–53
In or Out (Hadleigh), 133
Inquiry into the Human Mind, An (Reid), 13
insomnia, 4
Instagram, 119
Intacs (eye treatment), 119
International Olympics Committee, 60
Internet: celebrity halo effect on, 112, 118–20; cult solicitations on, 174; hypochondria triggers on, 28–29; reputable information on, 162; sexual content on, 122–23, 145; time perception and use of, 184–85. *See also* social media
interpretation, 12–15
intuition, 79, 80; animal, 192–93; defined, 187; failure to heed, 188–90, 191–92, 200; harnessing, 195–97, 211–12, 219–20; heeding, 187, 190–91; non-science-based, 188–93; overthinking and, 193–94; PI and, 187–88, 191–92, 196–97, 200, 202–3, 211–12, 219–20; science-based, 193–95
"Intuitive Investor, The" (Voss), 191
inventions, and intuition, 190
investing, and intuition, 190–91
Islam, 170–71

Jackson, Michael, 117
Jackson, Reggie, 85
James Cook University, 184
Japan, 136
Jenner, Caitlyn, 134–35
Jennings, Kerri Walsh, 78
Jennings, Waylon, 188–89
Jesus Cheeto, 88
Jesus sightings, 92
Jeter, Derek, 85
Jobs, Steve, 97
John, Elton, 133
Johnson, Lyndon B., 126
Jolie, Angelina, 117
Jones, Jim, 173
Jordan, Michael, 69, 76, 77, 80
Journal of Neuroscience, 194
Journal of Scientific Reports, 177
Judaism, 92

Kabat-Zinn, Jon, 24, 76–77
Kama Sutra, 124
Kardashian West, Kim, 113, 119–20
Kashi cereal, 161

Kentucky, 136
Kenya, 191–92
Keratoconus (eye disease), 2–3, 41–42,
    42n, 59, 78, 119
"kicking the bucket," 177–78
kinesthetic imagery, 56–57
King, Larry, 63, 158
King, Martin Luther, Jr., 97
King of Comedy, The (film; 1982), 114–15
Kinsey, Alfred, 135–36
Kolber, Suzy, 83
Konstantinovsky Palace (St. Petersburg,
    Russia), 65
Kopi Luwak coffee, 139–40
Koufax, Sandy, 81, 158
Kraft, Robert, 65–66
Krauss, Lawrence, 5
Kübler-Ross, Elisabeth, 50–51

Lady Chatterley's Lover (Lawrence), 126
Lady Gaga, 133
Lander, David, 22
Lane, Diane, 131
Lane, Nathan, 133
Lapine, James, 39
LASIK surgery, 118
laughter, 23
Lawrence, D.H., 126
Leach, Penelope, 179
Lear, Norman, 195
Lennon, John, 108n, 114, 196
Lennon, Sean, 196
lesbianism, 133, 134, 179
Lewis, Jerry, 114–15
Life after Life (Moody), 45
life coach, 194
Loch Ness Monster encounters, 90–91
London (England), 124
Long, Jeffrey, 45–46
Lopez, Jennifer, 114
Los Angeles Dodgers, 81–82
Lotto, Beau, 43
Louis C.K. (comedian), 116
luck, bad, 193
"Lucy in the Sky with Diamonds" (Beat-
    les), 196
Lufityanto, Galang, 194
luminance, 38
Lynde, Paul, 133

Machu Picchu (Peru), 36–37
macular degeneration, 3

Madame Bovary (Flaubert), 126
Maddow, Rachel, 133
Madoff, Bernie, 158–60
Madonna (singer/actress), 97
magic, 31
Malamud, Bernard, 76
Malkovich, John, 158
Manassas (VA), 109
Mantle, Mickey, 85
Marie, Scheana, 118
Mark Twain on Masturbation (Twain),
    129–30
Martin, George, 195
Martinez, Olivier, 131
mass hallucinations, 1
mass suicides, 173
masturbation, 121–22, 124–26, 200
Matrix, The (film; 1999), 10–11, 16
Maven, Max, 31
McCarthy, Jenny, 113
McCartney, Paul, 88–89, 90
McGwire, Mark, 82, 83
McKellen, Ian, 133
McLean, Don, 188
McLoughlin, Aoife, 184
meditation, 23–24, 76–77
Meek, Joe, 189
memory/memories: of amputated limbs,
    27; dark lucid dreams and, 35; physical
    exercise and improvement of, 72;
    self-visualization and improvement
    of, 75; time perception and, 182–83
mentalism, 31, 206, 212
Mercy Medical Center (Springfield, MA),
    90
messaging, 19
Mexico, 136
mind: brain and, 11; hypochondria and,
    21; NDEs and, 51–52; senses and, 14
mind-body connection, 22
Mindful Athlete, The (Mumford), 76–77
mindfulness, 23–25, 194
mindfulness-based stress reduction
    (MBSR), 24
Mindsight (Siegel), 24
Minersville (PA), 90
misperceptions, 9–10
Miss Universe Pageant (Moscow, Russia),
    63
Mistresses (TV series), 131
moment, living in the, 183
Monaco, 181

Montreal Canadiens, 80, 84
Moody, Raymond, 45
morality, 121
Morality in Media, 126
Moscow (Russia), 62, 63
Moses sightings, 92
Moss, Elisabeth, 172
Mother Teresa cinnamon bun, 88
Mourning Bride, The (Congreve), 109
movies, 122, 126, 131, 142n
MSG, 161
multiple sclerosis (MS), xv–xvii, 21–22
Mumford, George, 76–77
Munch, Edvard, 40
Murphy, Eddie, 116
Murray, Bill, 183–84
music, 39, 126–27
mysteries, unexplained, 36–37

Namath, Joe, 82–83
narcissism, 174
National Aeronautics and Space Administration (NASA), 178
National Basketball Association (NBA), 76–77
National Enquirer, 118
National UFO Reporting Center, 1
Natural, The (film; 1984), 76
Navratilova, Martina, 79–80
Neal, Mary, 45
Near Death Experience Research Foundation (NDERF), 46
near-death experiences (NDEs): accounts of, 46–48; of cardiac arrest survivors, 49–50; of celebrities, 53–54; common scenarios, 47; cultural influence on, 50–51, 52–53; mind and imagery of, 51–52; OBEs vs., 55; PI and, 52–54; statistics regarding, 46
negativity, 22, 24, 26, 71
neocortex, 16
nervous system, 24
Nessie encounters, 90–91
Nettles, Bonnie, 173
neuroscience, 12, 16–17, 38, 49, 50, 52
New England Patriots, 65, 85
New York Jets, 82–83
New York Mets, 81–82
New York Times, 18, 47
New York Yankees, 80, 84–85
Nimeiry, Gaafar, 170
North by Northwest (film; 1959), 160

novelty items, 144–45
"nun on a bun," 88

Oakland Raiders, 84
Obama, Barack, 61
Odom, Lamar, 82
Office, The (TV series), 214
Ohio State University, 18
O'Keeffe, Georgia, 97
Olympic Games (Rio de Janeiro, Brazil; 2016), 64
Olympic Games (Sochi, Russia; 2014), 59, 60–61, 63–65
Olympios, Corinne, 118–19
onanism, 124–25, 129–30. *See also* masturbation
On Death and Dying (Kübler-Ross), 50
On Grief and Grieving (Kübler-Ross), 50
Opening of Misty Beethoven, The (film; 1976), 128–29
optimism, 25–26
organic foods, and halo effect, 113–14, 200
organoid, 18
Osbourne, Ozzy, 54
Our Lady of the Fence Posts, 89
out-of-body experiences (OBEs), 1–2, 48, 55–56
overthinking, 193–94

Paller, Ken, 57
paranormal, belief in, 92
pareidolia, 90–92, 95
Patient TN, 57–58
"Paul is dead" hoax, 90
Payne, Charles, 134
Peale, Norman Vincent, 173
Pearson, Joel, 194–95
Penn & Teller: Bullshit!, 90
Penn and Teller (magic act), 31
People's Temple (cult), 173
perception: brain and, 2–4, 11–12, 18–19; defined, 12; guilt and manipulation of, 97; of illness, xvii–xix, 22–29; illusion as distortion of, 26–27; interpretation and, 12–15; malleability of, 43; misperceptions, 9–10; physical exercise and sharpening of, 72; reality vs., 4–6; religion and, 92–94; self-confidence and, 73–74; self-perception, 60, 61, 62–63, 66–68, 82–83; self-visualization and sharpening of, 75; of time, 176–78, 179–80, 182

perceptual contrast, 104
Perceptual Intelligence (PI): afterlife
    beliefs and, 45–46, 51; art and, 38–41;
    assessing, 200–210; celebrities and,
    111–15, 202; death and, 48, 50; defined,
    6, 12; distorted, 59, 67–68, 90, 199; as
    dynamic process, 124, 178; fanaticism
    and, 165–68, 169, 171–72, 174, 200, 201;
    hallmarks of, 93; hallucinations and,
    95; high, 6–7, 8, 22, 23–24, 26, 51, 58,
    73, 77, 78–79 (*see also specific topic*);
    hypochondria and, 29, 199; illusion
    and, 42–43; immaculate perception
    and, 88–89, 96–97, 203; improving,
    199–200, 210–17; intuition and, 187–88,
    191–92, 196–97, 200, 202–3, 211–12, 219–
    20; as learned skill, 7–8; life as test of,
    210; low, 8, 9–10, 13, 24, 28, 32, 36, 40,
    50, 51, 53, 58, 59, 70, 199–200 (*see also
    specific topic*); lowering of opponents',
    78; mass influence on, 28; NDEs and,
    52–54; OBEs and, 55; pareidolia and,
    90, 91; physical exercise and, 72–73;
    principles of, 13; reciprocity and, 99–
    101, 104, 106–7, 110, 199; revenge and,
    107, 108; sense of reality controlled by,
    17, 46, 58, 210–11; sexuality and, 121–22,
    123, 126–30, 132, 134, 135, 136, 200; social
    influence and, 152, 153, 158–59, 160,
    161, 163, 200, 204, 214–15; spending
    habits and, 141–42, 143–44, 145, 146–47,
    148–49; in sports, 69–70, 72–74, 82, 111;
    of sports fans, 85, 111; time and, 176,
    177, 178, 180–82, 183–84, 185, 200, 208;
    unexplained mysteries and, 37; willful
    lowering of, 31
Perceptual Intelligence (PI) Assessment,
    201–10
Perry, Tyler, 131
Phaedo (Plato), 41
Pham, Tommy, 78
Philadelphia Athletics, 80
Philadelphia 76ers, 81
Picasso, Pablo, 37, 97, 196
Pierce, David Hyde, 133
Pitt, Brad, 117
Pittsburgh Pirates, 80
Pittsburgh Steelers, 84
Plato, 41, 89
play, reciprocal vs. parallel, 103
Playgirl, 134

Play Misty for Me (film; 1971), 109
politics, 103–4, 201–2
Pompeii (Italy), 123
Ponzo, Mario, 42
pork barrel, 103
pornography, 122–24, 126–27, 128–29, 136,
    142n, 145
Portnoy's Complaint (Roth), 128
positivity, 23
preachers, 174
premonitions, 202–3
Presley, Elvis, 90
Preston, Kelly, 172
Prince of Peace Pierogi, 88
Principal, Victoria, 25
procrastination, 175–76
Proof of Heaven (Alexander), 47–48
prostate cancer, 126
prostitution, 124
Protestantism, 50–51
Proust, Marcel, 28
Pryor, Richard, 22, 116
Psychology Dictionary, 56–57
Puritans, 124
Pussy Riot, 62
Putin, Vladimir: authoritarian leadership
    style of, 60, 62; distorted PI of, 59, 199;
    Kraft Super Bowl ring and, 65–66;
    Russian obeisance to, 61–62; self-
    image of, 60, 61, 62–63, 66–68; Sochi
    Olympics (2014) and, 60–61, 63–65;
    Trump and, 62–63

quantum mechanics theory, 14
Quintessential Grilled Cheese Sandwich,
    87

radio, 112
Randi, James, 32
rape, 121
rapid eye movement (REM), 50
Raye, Martha, 112
Reagan, Ronald, 126
reality: false, 22; as illusion, 1, 2; of own-
    ership, 142–43; perception vs., 4–6; PI
    and, 17, 46, 58, 210–11; senses and, 16
reality TV, 118
receptor cells, 15
reciprocation principle, 99–100
reciprocity: altruism and, 102–3; com-
    promise/concessions in, 103–4; gender

differences in, 108–10; gifts and, 99–100, 105–7; guilt and, 101–2, 104; PI and, 99–101, 104, 106–7, 110, 199; revenge and, 107–10
Reed, Robert, 133
Reeves, Keanu, 10–11
regrets, 194
Reid, Thomas, 13
religion, 92–94, 124–26, 165, 168–71
remote viewing, 32
Republican (Springfield, MA, newspaper), 81
revenge, 107 10
rhythm, 79
Rice Krispies, 160
Richardson, J. P. ("Big Bopper"), 188–90, 200
Rickles, Don, 115
Rio de Janeiro (Brazil), 64
Rivera, Mariano, 85
Rock, Kid, 157–58
Rocky (film; 1976), 69
Rodchenkov, Grigory, 64
Roddenberry, Gene, 97
Rogaine, 160, 161
Rolling Stones, 180
Romans, ancient, 123, 124
Romeo and Juliet (Shakespeare), 141–42
Roosevelt, Theodore, 67
Rose, Pete, 82, 83
Roswell (NM), 34
Roth, Philip, 128
Russia: anti-doping penalties levied against, 64; bestiality in, 136; Olympic Games in (Sochi; 2014), 60–61, 63–65; PI in, 61, 62, 66; strong leadership style preferred in, 61–62, 66–67
Russian Popular Front, 62
Ruth, Babe, 85

Sagan, Carl, 34, 97
Sandoval, Tom, 118
San Francisco Examiner, 157–58
San Jose (Brazil) Faculty of Medicine, 177
Sargent, Dick, 133
Scandal (TV series), 131
schizophrenia, 94
Schumer, Amy, 116
Schwarzenegger, Arnold, 128, 129
science, 178
Scientific American, 23, 181

Scientology, 172–73
Scorsese, Martin, 114–15
Scott, Randolph, 133
Scream, The (Munch), 40
sea monkeys, 144–45
Seaver, Tom, 82
"Secret Life of Walter Mitty, The" (Thurber), 95
Seinfeld, Jerry, 84
self-confidence, 73–74, 80
"selfie stars," 119–20
selflessness, 102–3
self-perception, 60, 61, 62–63, 66–68, 82–83
self-visualization, 74–75, 76
Seljuk Turks, 169
sensation, 13
senses: brain and tricking of, 27; illusion and, 26–27; input from interpreted, 12–15, 16–17, 18–19; organs as overseers of, 15; reality and, 16
Serendipity 3 restaurant (New York, NY), 87
Seurat, Georges-Pierre, 39
sexual deviances, 121–22
sexuality: bondage, 129; Christian repression of, 124–26; homosexuality, 132–34, 178–79; infidelity, 121–22, 130–32; Internet and, 122–23; masturbation, 121–22, 124–26, 127–30, 200; PI and, 121–22, 123, 126–30, 132, 134, 135, 136, 200; pornography, 122–24, 126–27, 128–29; transgenderism, 134–35
sexual preferences, 121–22, 132–34, 178–79
Sha Na Na, 158
Shankar, Ravi, 158
Sharia law, 170
Shatner, William, 112
Shields, Brooke, 112, 113
"Shower Jesus," 90
Simpson, Pam, 75, 78, 79–80
Siegel, Daniel, 24
silver, perceived value of, 140–41
Silverman, Sarah, 116
Simpson, Jessica, 114
Simpson, O. J., 82, 110
Sinclair, John, 154
skepticism, 32–33, 37, 93, 97, 146
sleep deprivation, 4
sleep paralysis, 34–36, 37
Sly and the Family Stone, 156–57

smartphones, 184–85
smiling, 23
Snapchat, 119
Snopes.com, 214
Sochi (Russia), 59, 60–61, 63–65
Sochi National Park (Russia), 64
"Social Defaults: Observed Choices
    Become Choice Defaults" (Huh,
    Vosgerau, and Morewedge), 153
social influence: advertising and, 153,
    160–63; herd mentality and, 151, 153,
    200; lifetime pervasiveness of, 151–52;
    PI and, 152, 153, 158–59, 160, 161, 163,
    200, 204, 214–15; positive vs. negative,
    154–60; uncertainty and, 152–53, 159
social media: celebrity halo effect in, 111,
    112, 113, 118–20; fake news on, 203–4,
    213–14; health scares reported in, 29;
    reciprocity in, 99. See also Internet
social proof, 159
"Some Thoughts on the Science of
    Onanism" (speech; Twain), 129–30
Sondheim, Stephen, 39
Soros, George, 190–91
soul, 46, 51–52
South America, 136
souvenirs, 148–50
Spears, Britney, 88
spending habits: advertising and, 144,
    146–47; buyer's remorse and, 143; gag
    gifts/novelty items, 144–47; holiday
    gifts, 142–43; PI and, 141–42, 143–44,
    145, 146–47, 148–49; rarity perception
    and, 139–42; during vacations, 147–50
Spielberg, Steven, 97, 158
spirits, 36
Spiritual Brain, The (Beauregard and
    O'Leary), 45
Splenda, 160
sports: fallen stars in, 82–84, 111; fan fanat-
    icism in, 84–86, 111, 199, 201; PI in, 68,
    69–70, 72–74, 82, 111; self-confidence in,
    73–74; self-perception in, 82–83; self-
    visualization in, 74–75, 76; teamwork
    in, 79–82; the zone in, 76–79
stalking/stalkers, 114–15
Stanford University, 102
Stanwyck, Barbara, 133
StarGate Project, 32
Star Trek (TV series), 35

St. Louis Cardinals, 83, 85
St. Mary's Cathedral (Rathkeale, Ireland),
    1–2, 89
Stone, Sharon, 54
stress reduction, 22, 24
Sudanese Civil War, Second (1983–2005),
    170
suicides, mass, 173
Sunday Afternoon on the Island of La
    Grand Jatte, A (Seurat), 39
Sunday in the Park with George (Sond-
    heim and Lapine), 39
sunglasses, UV protection claims about,
    163n
Super Bowl, 65–66, 201
superhero films/TV, 33, 96–97
superpowers, illusions of, 144–45, 206, 212
survival mechanisms, 107
Sutherland, Donald, 54
Swann, Ingo, 32
Sydney (Australia), 89
Sykes, Wanda, 133
synergy, 79

tabloids, 117, 118
Takei, George, 133
Taylor, Elizabeth, 117
Taylor, Lawrence, 82
teamwork, 79–82
technology, 184–85
TEDx Talks, 42n
televangelists, 174
television, 112–13, 118, 122, 126, 130–31, 179
Temptation (film; 2013), 131
terrorism, 165, 170–71
Tesla, Nikolai, 190
testosterone, 108–9
Texas, 136
Thailand, 136
thalamus, 15–16
Thatcher, Margaret, 32
theater, 39–40
three-dimensional thinking, 38, 77
Thurber, James, 95, 97
time: animal perception of, 181; beliefs
    and, 178–79, 209, 215; clichés about,
    180, 182; films about, 183–84; mem-
    ory and, 182–83; perceptions of,
    176–78, 179–80, 182; PI and, 176, 177,
    178, 180–82, 183–84, 185, 200, 208;

procrastination, 175–76; technology and, 184–85
TN (blind patient), 57–58
To Heaven and Back (Neal), 45
Tomlin, Lily, 133
Townshend, Pete, 154–55
Tragically Hip, 158
transgenderism, 134–35
Travolta, John, 118, 172
Trump, Donald J., 62–63
trust, 93
Twain, Mark, 129–30
Twitter, 113, 119
Tyson, Mike, 82, 83

Ukraine, 60
uncertainty, 152–53, 159
UNESCO World Heritage sites, 64
Unfaithful (film; 2002), 131–32
unidentified flying objects (UFOs), 1–2, 34, 90
United Kingdom, 84
United Russia (political party), 62
United States: bestiality in, 136; brain research in, 18; dry eye disease in, 25; Internet porn viewing in, 122; marital infidelity rates in, 122; MS in, 22; NDEs in, 46; sexual misconceptions in, 124–26, 125n; sports fan fanaticism in, 84
University of Birmingham, 177
University of California, Los Angeles, 43
University of Connecticut women's basketball team, 80–81
University of Kentucky, 50
University of Michigan, 49, 50
University of New South Wales, 194–95
University of Ottawa, 56
University of Southern California, 194
Urban II, Pope, 168–69
User's Manual for the Brain, The (Hall and Bodenhamer), 6
UV protection claims, 163n

vacation spending, 147–50
Val Camonica (Italy), 135
Valens, Ritchie, 188, 189, 200
Vanderpump Rules (reality TV show), 118
van Gogh, Vincent, 40–41

Vanity Fair, 134
Vatsyayana (Hindu philosopher), 124
Vick, Michael, 82, 83
Victorian Era, 124
Virgin Mary, visions of, 1–2, 50–51, 87–88, 89–90, 93–94, 96–97
visual cortex, 42–43, 56–57
visualization, 40. *See also* self-visualization
Voevoda, Alexey, 64
vomit-ins, 165
Voss, Jason, 191

Wachowski, Andy, 10
Wachowski, Larry, 10
Walker, Clay, 22
Warhol, Andy, 28
wedding gifts, 105–6
Weill, Sandy, 65
Westheimer, Ruth, 127
Westworld (film; 1973), 130
Westworld (TV series), 130–31
Who, The, 154
Wiesel, Elie, 158
Wilde, Oscar, 133, 178
Williams, Montel, xv–xx, 21–22, 26
Williams, Robin, 116
Williams, Tennessee, 28
Wills, Jerry, 36–37
Winfrey, Oprah, 97, 117, 118, 190
Woods, Tiger, 82
Woodstock Festival, 154–57
workplace, nonindividual thinking at, 214
World Anti-Doping Agency, 64
World Series (2004), 85

X-Files, The (TV series), 34
X-Ray Specs, 145

Yanek, Lisa R., 23
Youngman, Henny, 116
YouTube, 42n, 119, 157

Zamalloa, Ana, 36–37
zone, getting into the, 76–79
zoophilia, 136
zoo porn, 136
Zubkov, Alexandr, 64